CRASH COURSE ON PYTHON SCRIPTING FOR ABAQUS

Learn to write python scripts for ABAQUS in 10 days

Renganathan Sekar

Disclaimers

The author is not responsible for the accuracy of the python scripts published in this book. This book is not endorsed by developers of ABAQUS and the author has no relationship with them. This book has been written with an intention to teach people who want to learn Python scripting for ABAQUS. The author is not liable to any damages caused directly or indirectly because of the instructions given in this book.

No part of this book can be reproduced or transmitted in any format by any means – Electronic, mechanical, photocopying, recording, etc. without the prior written consent of the author. Kindly please refrain from publishing or uploading this book in Internet.

As this book is self-published, please contact the author directly for any questions or group discounts or suggestions – renganathan.sekar@gmail.com. Photo used in the book's cover – Esther Jiao, Unsplash

Printed by CreateSpace, An Amazon.com Company

ISBN-13: 978-1724801319

ISBN-10: 1724801317

Copyright © 2018 Renganathan Sekar

All rights reserved.

DEDICATION

To the lotus feet of my Guru.

ध्यानमूलं गुरुमूर्तिः पूजामूलं गुरुर्पदम् ।
मन्त्रमूलं गुरुर्वाक्यं मोक्षमूलं गुरूर्कृपा ॥

Why this book?

1. To understand and put in place efficient python scripting techniques for ABAQUS. And develop a practical perspective of the application of Finite Element Method (FEM).

2. If you have already started Python scripting for ABAQUS, you would have noticed the scarcity of step by step tutorial documents. This book is a sincere attempt to address this trivial concern. I have provided well-commented python scripts from 9 different categories to get you started.

3. When it comes to learning a new skill, you might feel confused due to a myriad of options. Some of the common pitfalls would be not choosing the right resources. Or the lack of a strategic learning approach. Add to this, the dangerous "starting troubles". This phase is trickier and you have to cross this to do your task. I present this book in such a way that the reader can maneuver these inevitable pitfalls. And actually start writing scripts that work.

4. Start following the practices in this book. And I am sure that you can notice a significant increase in your productivity. Python scripts give you the leverage to automate repetitive tasks in FEM simulations using ABAQUS. Who doesn't like to go home earlier?

5. Practice the techniques in this book long enough. You will adopt a certain mentality while approaching newer problems. These techniques will become a mental framework to help you reduce trivial errors in FEM simulations. And help focus your energies on actual problem solving.

6. Time is precious. Period. I will straight away jump in to the problems, emphasizing on the solution methods. My intention is to make you write efficient python scripts in ten days. If you practice one chapter per day with this book. Of course, for a deeper understanding of FEM Theory or Python language, you will have to refer standard text books.

Have I convinced you enough? Let's get started.

Who will get the most out of this book?

Beginners new to ABAQUS/Python scripting environment will find this book very valuable. Follow the examples and practices in this book daily. And you can solve bigger problems from the industry. With confidence and clarity of what you are doing. Of course, you get refined by consistent practice and experience. But as clichéd it may sound, if you have a strong foundation, its easy to build on it. Compounding works great with knowledge too.

Academics can plan their syllabus or coursework based on this book. I have a Master's degree in computational sciences from RWTH Aachen University, Germany. And I work in an academic-industry setting in South Korea. This has given me insights on the typical course structure in top universities. Based on these inputs, I have tailored the contents of the book so that you can use it for coursework.

Graduate students working under tight deadlines will definitely find this book very useful. One of the important requirements from the perspective of your employer would be the re usability of your FE method. It makes a lot of sense for them too as they can extend the scope of your work beyond your tenure. Using python scripting for ABAQUS can help you do that exactly with efficiency. And I do have a request for you, dear graduate students. I know the power of 'word of mouth'. So if you found this book to be useful, please recommend this book to your friends.

For experienced FEM analysts, this book might be a refresher in some aspects.

Authors' note:

I prefer using direct voice to address the readers and have done this in most places throughout the book. So the writing style might resemble a technical blog. This is a conscious attempt to create a connect with my readers. I wish this does not hinder your understanding of the subject matter.

The ideal way to use this book

If you want to learn the skill of writing Python scripts for ABAQUS in ten days or even less, please do the following.

- ❖ I am a big believer of habits as consistent efforts are always very good and important for the long run. You will have to promise me that you will finish one chapter per day. This is a sensible target even if you are a stranger to Python scripting.

- ❖ You never have to memorize the syntax of statements or methods. ABAQUS Python scripting manual has detailed explanations for everything. In this book, you will learn the workflow of Python scripting and the process of writing scripts in general. Of course with examples. Always have the scripting documentation by your side whenever you write scripts. This boosts your understanding.

- ❖ Of course, you can copy and paste the scripts from this book. I have tested them many times and it will work. No doubts about that. But when you are trying to learn the art of scripting, please type them on your own and make minor changes. You will definitely make some mistakes along this process. But this will reinforce your learning and make you even stronger.

A general tip:

If you are writing a script and get struck up somewhere, try doing the same thing using the GUI. And record what you do using a Macros in ABAQUS. This is like Excel macros. After finishing the sequence of actions, you can have a look into the macros file. This file is present in the ABAQUS working directory. And it would have the list of equivalent Python statements for the actions that you performed in the GUI.

Contents

DEDICATION iii

Why this book? iv

Who will get the most out of this book? v

The ideal way to use this book vi

Acknowledgements x

1. Introduction to ABAQUS-Python scripting 11

1.1 Basics of FEM – 7 general steps 12

1.2 Python from ABAQUS perspective 14

    1.2.1 Adding comments 14

    1.2.2 Variables and their correct usage 15

    1.2.3 Accessing/organizing elements from lists 16

    1.2.4 Conditional statements 18

    1.2.5 Dictionaries 20

    1.2.6 Enabling user input 20

    1.2.7 How to run a Python script in ABAQUS 21

1.3 General workflow in python scripts 22

1.4 Unit system 22

2. Static analysis of an overhead hoist 24

2.1 Problem description 25

2.2 Python script 26

2.3 Summary	33
3. Deflection of a cantilever beam	34
3.1 Problem description	35
3.2 Python script	35
3.3 Summary	42
4. Static / Dynamic analysis of a connecting lug	43
4.1 Problem description	44
4.2 Python script	45
4.3 Summary	52
5. Bending analysis of a plate	53
5.1 Problem description	54
5.2 Python script	54
5.3 Summary	61
6. Contact analysis of an electrical switch	62
6.1 Problem description	63
6.2 Python script	64
6.3 Summary	71
7. Wireframe analysis(3D) with box profile	72
7.1 Problem description	73
7.2 Python script	74
7.3 Summary	85
8. Steady state thermal analysis	86
8.1 Problem description	87

8.2 Python script	88
8.3 Summary	97
9. Script for sending email after job completion	98
9.1 Problem description	99
9.2 Python script	100
9.3 Summary	108
10. Parametrization of a truss	109
10.1 Problem description	110
10.2 Python script	111
10.3 Summary	118
References	119
About the author	119

Acknowledgements

प्रकृतेः क्रियमाणानि गुणैः कर्माणि सर्वशः |
अहङ्कारविमूढात्मा कर्ताहमिति मन्यते ||

I am grateful to my parents for giving me the freedom to choose what I wanted in life and then trust me completely. Thanking you in words will never be enough though. I hope to repay your kindness by noble actions.

Thank you very much, my dear wife Meenakshi for your support throughout the course of writing this book. I am blessed to have you in my life.

Special thanks to my friend Gokulnath for taking the time to proofread the book and give suggestions despite his busy schedule.

1. Introduction to ABAQUS-Python scripting

Things to learn

- 7 general steps of any FEM simulation
- Basics of Python from ABAQUS perspective
- General workflow in a Python script

1.1 Basics of FEM – 7 general steps

Before writing python scripts, it is good to have a quick refresher on the basics of FEM. We will do that in this section.

When I first saw some FEM simulation results 10 years ago, it was fascinating on the one hand. And intimidating on the other. Having spent some time now, it is not intimidating anymore. But the fascination part keeps growing as I keep learning new things. So, I will share the general workflow that I keep at the back of my mind when I am solving a FEM simulation.

There are some mandatory things in FEA, regardless of the problem type or its complexity. As a beginner, if you have a clear idea of the workflow of a FEM simulation, it is easy to grasp things. And solve problems in style. Knowing the workflow is important from the point of validating your results as well.

If you have used some software even at a basic level, you will get some contour plots based on the inputs you give. They say this: Garbage-In-Garbage-Out. So in any FEM software, getting an answer always doesn't mean that it is the right one. So, you must be aware of your input data and the sequence of operations to tackle unexpected problems. And then get the desired simulation results.

Are you ready to know the 7 general steps in a FEM simulation?

Let's get started.

Step 1: Modelling

The part is modeled without complicated geometrical features. This is the first and the most crucial step in any analysis. This step, if done right, will save a lot of time. So, pay a lot of attention to the geometry. Also try to understand the reason why you are simulating this component. This will provide you the insights to remove insignificant features from your geometry. And save some computational time and unwanted complexity. Always remember, simpler, the better.

Step 2: Material definition

You have to define the material properties in this step. These material properties depend on the analysis type. Literature survey for similar problems could help you a lot in tackling this step. Luckily, If your material already exists in the material library of your FE tool, your job is almost done. Select and import the material properties from the library. As simple as that. Otherwise, you have to define the constitutive equation. And the corresponding material parameters based on the material behavior.

Step 3: Definition of loads

This step is about the definition of external forces acting on the part or the body force under the component weight. You have to be careful about the force definition type in order to avoid encountering problems like singularities.

Step 4: Boundary conditions

This step is done to reduce the complexity of the problem from an engineering sense. For example, in some problems, the user might be knowing some initial conditions before starting an analysis (like displacement of a point in the geometry). Such cases fall into the "Initial value problem" category.

Step 5: Meshing

Your geometry is divided into smaller and simpler shapes called as finite elements. Any standard FEM textbook will have the information about the different types of elements and their applications. Recently, a lot of software tend to automate this process avoiding the trouble for the end user.

Step 6: Solution

This step happens in the backend after the definition of the simulation properties. This is called as discretization. In simple terms, it means that the partial differential equations are converted into algebraic equations. Doing this helps the FE code to represent equations in terms of matrices. Matrices of individual elements are assembled into global matrices for the entire geometry which is then solved by FE solvers for unknown variables.

Step 7: Post-processing

Any FEM software will have an indicator to show the user if the analysis has succeeded. Once the solver gives this message, the desired results can be visualized through contour plots or graphs. For beginners, this would be the exciting part. A decent physical understanding of the phenomenon that you are simulating is crucial. It's always much better to validate your results with literature or experiments.

These 7 steps are very basic and generic in nature. With these steps in mind, let's progress into learning the basics of Python scripting. Of course from the perspective of using it for carrying out FE simulations using ABAQUS.

1.2 Python from ABAQUS perspective

I will now introduce the concepts of Python which are relevant to ABAQUS environment. My intention is to make you write efficient Python scripts in the shortest possible time. So I am covering only the things which are of absolute necessity. As you progress through the book, I am confident that you will pick up the usage of the Python language. And, If you want an in-depth understanding of the Python language, (Matthes, 2016) can be a good starting point.

1.2.1 Adding comments

This might sound trivial to some readers but this takes top priority always, irrespective of the programming language. It is always a good practice to code with self-explanatory comments. This ensures that the people who might use your code in the future find it easier to make changes without spending a lot of time to understand. If proper commenting is not done, even you can find it difficult to understand your own code. A # symbol is used to write comments in a Python code. Here are some examples of comments.

```
# This is my first comment.

# Commenting is actually good in coding!
```

Make sure that you add useful and self-explanatory comments. You will see a lot of comments in my scripts in the coming chapters, which will hopefully help you understand the code better.

1.2.2 Variables and their correct usage

You will be using a lot of variables from now on. If proper variable names are used, it improves the readability of your code. So, let's look into some rules that can be kept in mind when you are using variables in Python.

Rule 1:

A variable name cannot start with a number. It can start with letters and then have numbers and underscores.

Correct: `vertices_for_force`   Incorrect: `1vertex`

Rule 2:

Avoid the usage of python keywords like `raise`, `class`, `def`, `with`, `yield`, `lambda`, etc.,

Rule 3:

Spaces are not allowed in a variable name. Use an underscore if you have two words in a variable name.

Correct: `vertices_for_force`   Incorrect: `vertices for force`

Rule 4:

Variables must be descriptive of what they represent. Avoid the usage of variable names with a single alphabet though it might be easy initially. These rules cover most of the mistakes that beginners usually make. During the course of the book, I will add on to this list whenever there is a need.

1.2.3 Accessing/organizing elements from lists

A list (similar to the more familiar term 'array') is a collection of items entered/created in a particular order. It can be integers, strings, objects or similar items. Example for a python list would be

`vertices_of_triangle` = ['point1', 'point2', 'point3']

As you can see from the example, square brackets indicate that it is a list and the individual list items are separated by a comma.

The position of the individual list item in the list is called as index. In Python language, indexing starts from 0.

Printing `vertices_of_triangle[0]` would return `point1`

Similarly `vertices_of_triangle[2]` would return `point3`

Length of the list:

This list contains 3 elements – which is the length of the list. This can be obtained by using the `len` statement.

`len[vertices_of_triangle]` would return 3

Accessing the last element:

If you want to access the last element of the list, you can ask for the item at index -1.

`vertices_of_triangle[-1]` would return `point3`

You might find it difficult to appreciate the importance of this feature through this example. But imagine a situation where the list has more than 15 items and you are not sure of the length of the list. In such cases, you can directly access the last element of the list by referring the list item at index -1.

Appending an element to the list:

By using the append() method, you can add new data to the existing list item. For example,

vertices_of_triangle = ['point1', 'point2', 'point3']

If you would like to add one more point to the triangle list, use

vertices_of_triangle.append('mid_vertex')

Now the list gets updated to

vertices_of_triangle = ['point1', 'point2', 'point3', 'mid_vertex']

Removing an element from the list:

The del statement can be used to remove an item from the list based on its index.

We can remove the 'mid_vertex' from the list by using

del vertices_of_triangle[3]

You have to remember that the value that was removed can no longer be accessed once you have used the del statement.

Sorting the list:

As the name suggests, sort() method is used for sorting the list items. Let's consider this example.

success = ['luck', 'hardwork', 'perseverance']

Now sorting out success means success.sort() which will return

['hardwork', 'luck', 'perseverance']

True with our life too, right?

The sort() method has sorted the list items in the alphabetical order.

Avoid index errors:

As we would be using index of the list items to refer them, please make sure that you are using the correct index. The length of the list is always greater than the highest index by one as the indexing starts from 0.

For loops:

Loops are unarguably, one of the most important concepts in programming. They are used to carry out repetitive tasks on each and every item of a list.

The for loops in a Python script are very easy to read and it's usage is similar to how we use statements in the English language.

Let's take this example.

```
icecream = ['milk', 'sugar', 'icecubes']

for ingredients in icecream:
    print(ingredients)
```

This would print all elements of list `icecream`.

There are two important things to notice here.

1. Observe the indentation of the third line of code where the ingredients are printed. This is purposefully done to include the statement into the loop. One of the common mistakes is to avoid the indentation or overuse it. Make sure that you indent the statements only that are required to loop.

2. There is a colon at the end of the for loop. This is also a must in all conditional statements as well as loops in Python.

1.2.4 Conditional statements

The conditional statements are self-explanatory in nature. I will provide you examples and add comments wherever necessary.

Again, I would like to stress on the indentation being used.

if and if-else Statements:

```
success = ['luck', 'hardwork', 'perseverance']

for item in success:

    if item == 'hardwork' :

        print (item.upper())

    else:

        print(item.title())
```

These statements search for the item 'hardwork' in the list 'success'. If the item is found, then the respective item is printed fully in uppercase.

If the item is not found, then it is returned in title case.

So for this example, the output, as you might have guessed would be:

HARDWORK

if-elif-else Chain:

Let's take the example of suggesting a suitable attire based on the weather. Ignore the units of temperature for now.

```
temperature = 30

if temperature < 20:

    print("Use a jacket")

elif temperature < 10:

    print("Use a scarf")

else:

    print("Wear loose and comfortable clothes")
```

Pay close attention to the indentation used in this example.

1.2.5 Dictionaries

Dictionaries in Python language allow to store and access information in a much organized format. With a proper understanding of this concept, we would be able to model a lot of real-world applications. These dictionaries are a collection of key and value pairs. Each key can be used to access its value in the dictionary. Consider some examples.

```
shapes_vertices = {'triangle': 3, 'rectangle': 4, 'pentagon': 5 }

instrument1 = {'Name': 'Flute', 'Holes': 7, 'Scale': 'A' }

person1 = {'Name': 'Human', 'Age': 25, 'Sex': 'Female'}
```

If a key is provided, then Python returns the corresponding value from the dictionary. As you can see, a dictionary is created inside {}.

1.2.6 Enabling user input

Enabling your script to get input from users is important as the scope of your program keeps getting wider. The input() function can be used to do this. When you invoke this function, Python pauses the execution of the program and waits for input entry from the user. Here is an example.

```
fact = input("Tell me how much you like reading this book so far")

print(fact)
```

So, the compiler waits for an entry from the user. Once the input is given, it prints out the message. I really hope you like reading this book so far ☺

By this time, you might have noticed that none of the Python statements end with a semicolon similar to some programming languages. This makes it easy to read.

1.2.7 How to run a Python script in ABAQUS

I have covered the basics. So it is time to learn how to run a Python script in ABAQUS from its GUI, which takes us into the topic of text editors.

Text editor:

You would be needing a text editor to type your Python scripts. This can be a simple Notepad too. The text file has to be saved with .py extension.

There are many popular text editors like Notepad++, UltraEdit, etc.,

My personal favorite is Microsoft Visual Studio Code. It is free and has a lot of useful features. You can choose a text editor based on your preferences. Make sure you have autocomplete options in the editor as it is important for productivity and also helps to avoid typographical errors.

Saving the text file with .py extension:

When you open a text editor, open a new text file, save it with .py extension and then start typing your code. Once you have completed fully, then save the file once again.

In ABAQUS/CAE GUI , go to **File > Run Script...**

This will ask you to open a Python script file. Now navigate to the folder where you have saved your Python script and open it.

ABAQUS executes your Python script and you can see the respective things happening on the screen as the code is getting executed. If there is an error, then the compiler prompts a dialog box with the details of the error and line in which the error has happened. This can be used to check your code and make changes if any.

1.3 General workflow in python scripts

Keep in mind, the workflow that we would be following for all simulations in ABAQUS. This framework is broad in nature and specific steps get added to the workflow based on the analysis type.

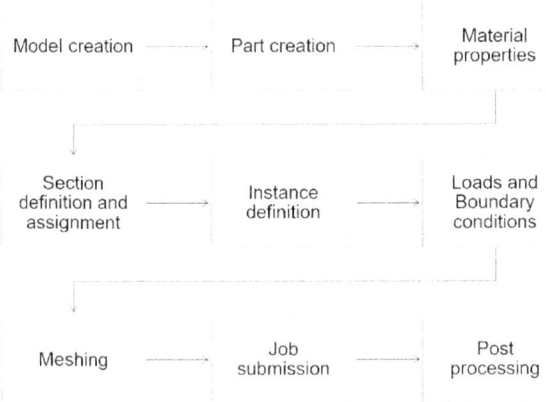

Figure 1.1 . Workflow in Python scripts

1.4 Unit system

This is also a very important topic that can never be ignored. Before you start writing the script, you must be clear with the unit systems, that you plan to use. As ABAQUS has no built in system of units, you have to make sure that you follow a consistent system of units throughout the simulation.

Quantity	SI	SI (mm)	US Unit (ft)	US Unit (inch)
Length	m	mm	ft	in
Force	N	N	lbf	lbf
Mass	kg	tonne (10^3 kg)	slug	lbf s^2 in
Time	s	s	s	s
Stress	Pa (N m^2)	MPa (N mm^2)	lbf ft^2	psi (lbf in^2)
Energy	J	mJ (10^{-3} J)	ft lbf	in lbf
Density	kg m^3	tonne mm^3	slug ft^3	lbf s^2 in^4

Figure 1.2 Consistent units (DS Simulia, 2008)

So far, I have introduced the bare minimum pre-requisites that can be used to write Python scripts in ABAQUS. Some of you might think that this introduction might not be sufficient. If you are one among them, you are partially correct as well. But my intention is not to make it look simple. I am a firm believer in learning by doing. So I have consciously given just the basics in this chapter and will teach you the necessary concepts, whenever necessary, in the coming chapters where we would actually write the scripts.

2. Static analysis of an overhead hoist

Things to learn

※ 'One main word' strategy for naming variables

※ Statements and methods used in a Python script

※ Writing a Python script to evaluate static deflection of truss

2.1 Problem description

This chapter would be about finding the static deflection of an overhead hoist structure. The structure is a truss which is constrained at the left end and mounted on rollers at the right end.

The example is taken from the ABAQUS user manual. Figure 2.1 shows the schematic of the overhead hoist. All members are circular steel rods, 5mm in diameter.

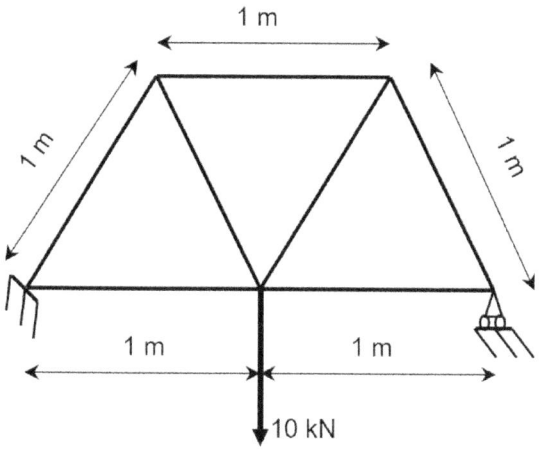

Properties	Values
Density	7800 kg/m^3
Young's modulus	200E9 Pa
Poisson's ratio	0.3

Figure 2.1 Schematic of the overhead hoist (DS Simulia, 2008)

A 10 kN load is applied on one of the nodes and the static deflection of the truss structure needs to be estimated. Once we will write the script for this task, we can use the same script for finding the dynamic response by making slight modifications (will be dealt later).

Let's get started.

2.2 Python script

For solving this problem, you will have to

- ➢ Sketch the truss
- ➢ Define the material
- ➢ Create a section and assign it
- ➢ Create an instance and step
- ➢ Define the boundary conditions and loads
- ➢ Mesh the part
- ➢ Create the job, submit it and carry out post-processing

As discussed in Chapter 1, you would have to perform the above steps for every analysis that you wish to do in ABAQUS. Having this work sequence in mind can help you write the script more efficiently.

'One main word' strategy:

In all the scripts throughout this book, I will be using an idea, which I would like to call as 'One main word' strategy. I will choose this 'One main word' based on the problem context and use it while naming the variables. You will appreciate this strategy as it will eliminate the need to keep thinking new variable names as well as avoid silly mistakes in the process.

For example, this example deals with an over hoist structure. So I would be naming variables in the likes of overhoistModel, overhoistPart, overhoistSection, overhoistMaterial, etc.,

As you can see, 'overhoist' is the one main word which I will be using throughout. You will understand this much better as you read the script. I have added comments into the code wherever required.

Repeating, comments are statements that begin with #.

```python
# Creating an analysis model of an overhead hoist


# The first three lines are required to import the required ABAQUS modules
# and create references to the objects that are defined by the module.
# The second line means that you are importing the symbolic constants
# (variables with a constant value) that have been defined by the scripting
# interface of Abaqus.It is a good practice to include these three lines at
# the top of every script that you write.The third line is used for acessing
# the objects of Region() method which is defined inside the regionToolset
# module.

from abaqus import *
from abaqusConstants import*
import regionToolset

# This line is required to make the ABAQUS viewport display nothing
session.viewports['Viewport: 1'].setValues(displayedObject=None)

# Model creation
# By default, ABAQUS creates a model named 'Model-1'. We will use the
# changeKey() method to change the name of the model. 'mdb' will giveaccess
# to the model database and we will assign it ot 'overhoistModel'.

mdb.models.changeKey(fromName='Model-1', toName='Overhoist')
overhoistModel = mdb.models['Overhoist']


# Part creation

# These two statements will provide access to all the objects related to
# sketch and part. Including this is not mandatory, but is a good practice.

import sketch
import part

# In this set of statements, we would define a sketch by using the
# ConstrainedSketch() method. This method in turn has Line() method which
# can be used to draw the lines that we want in the sketch. Enter a value
# for 'sheetSize' based on your overall problem size.
```

```
overhoistSketch = overhoistModel.ConstrainedSketch(name='overhoist sketch
2D', sheetSize=10.0)
overhoistSketch.Line(point1=(0,0), point2=(1,0))
overhoistSketch.Line(point1=(1,0), point2=(2,0))
overhoistSketch.Line(point1=(0,0), point2=(0.5,0.866))
overhoistSketch.Line(point1=(0.5,0.866), point2=(1.5,0.866))
overhoistSketch.Line(point1=(1.5,0.866), point2=(2,0))
overhoistSketch.Line(point1=(0.5,0.866), point2=(1,0))
overhoistSketch.Line(point1=(1,0), point2=(1.5,0.866))

# Using the sketch that we have created, let us create the part using the
# Part() method. TWO_D_PLANAR is an example of symbolic constant.
# The BaseWire() method is used to create a feature object based on the
# sketch we have created.

overhoistPart = overhoistModel.Part(name='overhoist',
dimensionality=TWO_D_PLANAR, type=DEFORMABLE_BODY)
overhoistPart.BaseWire(sketch=overhoistSketch)

# Material creation

# First get access to objects relating to materials by using the import
# statement. Define the name of the material, which would be used further
# in the script. The Density() and Elastic() objects are used to specify
the
# density and elastic properties as name suggests. The input arguments to
# Density() looks so as it is actually a table with density values with
# respect to temperature. Here we don't need that. So we use blank spaces.
# The same logic applies to Elastic() where it is a table with Young's
# modulus values with respect to Poissons's ratio

import material

overhoistMaterial = overhoistModel.Material(name='Overhoist Steel')
overhoistMaterial.Density(table=((7800, ),       ))
overhoistMaterial.Elastic(table=((200E9, 0.30),  ))
```

```
# Section creation and assignment

# Get access to the section objects by using the import statement. Then
# create a truss section by using the TrussSection() method. You can see
# that we have refered to the material we created in the last step.
# The next step is to assign the created section to truss members. For this
# we use the findAt() method to find the edges at the provided vertices of
# the part. With the edges, we can create a region which can be assigned to
# the created section

import section
overhoistSection = overhoistModel.TrussSection(name='Overhoist Section',
material='Overhoist Steel', area=1.963E-5)
overhoist_section_edges = overhoistPart.edges.findAt(((0.5, 0.0, 0.0),
),((1.5, 0.0, 0.0), ),((0.25, 0.433, 0.0), ),((1.0, 0.866, 0.0), ),((1.75,
0.433, 0.0), ),((0.75, 0.433, 0.0), ),((1.25, 0.433, 0.0), ))
overhoist_region = regionToolset.Region(edges=overhoist_section_edges)
overhoistPart.SectionAssignment(region=overhoist_region,
sectionName='Overhoist Section')


# Assembly creation

# Get access to the assembly objects by using the import statement. The
# rootAssembly is an assembly object which is a member of the Model object.
# Create an instance of the part by using the Instance() method. By default,
# the 'dependent' parameter is set to OFF. Set this to ON. We have already
# defined the part name as 'overhoistPart'. We will refer to that now.

import assembly
overhoistAssembly = overhoistModel.rootAssembly
overhoistInstance = overhoistAssembly.Instance(name='Overhoist Instance',
part=overhoistPart, dependent=ON)

# Step creation

# Get access to the step objects by using the import statement. The
# StaticStep() method is used to create a static step which could be used
# for loading. This is the step next to 'Initial' step created by default.
```

```python
import step
overhoistModel.StaticStep(name='Loading Step', previous='Initial',
description='Loads will be applied in this step')


# Definition of field output requests


# ABAQUS, by default creates field output requests and names it as
# 'F-Output-1'. We will change it's name as well as set the parameters that
# we are interested in investigating.

overhoistModel.fieldOutputRequests.changeKey(fromName='F-Output-1',
toName='Required Field Outputs')
overhoistModel.fieldOutputRequests['Required Field
Outputs'].setValues(variables=('S', 'U', 'RF', 'CF'))


# Apply loads


# A load of 1000N has to be applied at vertex (1.0, 0.0, 0.0). So we
# should identify it using the findAt() method. ConcentratedForce()
# method is used to apply the force of 1kN at this vertex. Note that, we
# have refered to the step that we created sometime back.

vertex_for_force = (1.0, 0.0, 0.0)
force_vertex = overhoistInstance.vertices.findAt((vertex_for_force,))
overhoistModel.ConcentratedForce(name='Force1', createStepName='Loading
Step', region=(force_vertex,), cf2=-1000.0, distributionType=UNIFORM)


# Apply Boundary conditions


# Similarly the boundary conditions have to be applied at the left and
# right end of the truss structure. So first we identify the vertices and
# then we apply the boundary conditions at these vertices.

vertex_coords_encastre = (0.0, 0.0, 0.0)
vertex_coords_rolling = (2.0, 0.0, 0.0)
vertices_for_encastre =
overhoistInstance.vertices.findAt((vertex_coords_encastre,))
```

```
vertices_for_rolling =
overhoistInstance.vertices.findAt((vertex_coords_rolling,))


# EncastreBC() is used to create a Encastre joint. The DisplacementBC()
# method is used to define the displacement behavior on the region. The
# symbolic constant 'SET' means that we have constrained the dofs. In this
# example, we have constrained the translational DOFs in the y direction.

overhoistModel.EncastreBC(name='EncastreBC', createStepName='Initial',
region=(vertices_for_encastre,))
overhoistModel.DisplacementBC(name='RollingjointBC',
createStepName='Initial', region=(vertices_for_rolling,), u1=UNSET, u2=SET,
ur3=UNSET, amplitude=UNSET, distributionType=UNIFORM)

# Mesh creation

# Get access to the mesh objects by using the import statement. We will use
# the predefined regions for element type definition and for seeding the
# edges. T2D2 is the 2d element type for truss elements. We define the
# mesh size by seeding the edges by a number. This number can be increased
# to have a finer mesh. generateMesh() method is used to mesh the part.

import mesh
element_type_for_mesh = mesh.ElemType(elemCode=T2D2, elemLibrary=STANDARD)
overhoistPart.setElementType(regions= overhoist_region,
elemTypes=(element_type_for_mesh, ))
overhoistPart.seedEdgeByNumber(edges= overhoist_section_edges, number=2)
overhoistPart.generateMesh()

# Job creation
# Get access to the job objects by using the import statement. The Job()
# method is used to create a job. Make sure that you enter the correct name
# of the model. Most of the arguments entered here are not mandatory. You
# can edit the values base don your requirements.

import job
mdb.Job(name='OverhoistAnalysisJob', model='Overhoist', type=ANALYSIS,
explicitPrecision=SINGLE, nodalOutputPrecision=SINGLE,
```

```
description='Analysis of an overhoist crane with concentrated loads',
parallelizationMethodExplicit=DOMAIN, multiprocessingMode=DEFAULT,
numDomains=1, userSubroutine='', numCpus=1, memory=50,
memoryUnits=PERCENTAGE, scratch='', echoPrint=OFF, modelPrint=OFF,
contactPrint=OFF, historyPrint=OFF)

# The submit() method is used for submitting the job for analysis. The
# waitForCompletion() makes ABAQUS wait till the job is fully executed.

mdb.jobs['OverhoistAnalysisJob'].submit(consistencyChecking=OFF)
mdb.jobs['OverhoistAnalysisJob'].waitForCompletion()

# Post processing
# Get access to the visualization objects by using the import statement.
# We save the odb object and path to variables which could be used for
# visualization. The node labels and element labels are turned on for
better # clarity.The viewport size can also be set.

import visualization

overhoistPath = 'OverhoistAnalysisJob.odb'
odb_object = session.openOdb(name= overhoistPath)
session.viewports['Viewport: 1'].setValues(displayedObject=odb_object)
session.viewports['Viewport:
1'].odbDisplay.display.setValues(plotState=(DEFORMED, ))
overhoist_deformed_viewport = session.Viewport(name='Overhoist in Deformed
State')
overhoist_deformed_viewport.setValues(displayedObject=odb_object)
overhoist_deformed_viewport.odbDisplay.display.setValues(plotState=(UNDEFOR
MED, DEFORMED, ))
overhoist_deformed_viewport.odbDisplay.commonOptions.setValues(nodeLabels=O
N)
overhoist_deformed_viewport.odbDisplay.commonOptions.setValues(elemLabels=O
N)
overhoist_deformed_viewport.setValues(origin=(0.0, 0.0), width=150,
height=150)
```

2.3 Summary

Do not worry if you don't understand every line in the script. It is natural at this point of learning to write Python scripts for ABAQUS. Do not panic!

As of now, try to make mental connections. Between the input arguments that we used in the script and the ones that we define in the ABAQUS CAE Graphical User Interface(GUI).

Some of the statements in this script will be used throughout the course of this book. So when you finish reading all the chapters in this book, you will get a clear picture of the scripting practice. And start writing scripts on your own.

You can also copy and paste some parts of the this script into your script and make minor modifications. For example, from the material definition or from the Postprocessing section. Or for generating similar output reports every time after you complete a simulation. This will save you a lot of time if you have to repeatedly use the same material for many simulations. .

To summarize, in this chapter we learnt how to write a Python script for finding out the static response of an overhead hoist with a point load of 1kN. You also learnt to visualize the simulated results after successful job completion.

3. Deflection of a cantilever beam

Things to learn

※ Statements and methods used in a Python script

※ Writing a Python script to evaluate static deflection of a cantilever beam

3.1 Problem description

In this chapter, we will find out the static deflection of a cantilever beam by writing a Python script. The cantilever beam under investigation is shown in Figure 3.1, The steel beam has a rectangular cross section and is subjected to an uniform pressure load on its top surface. Like any other cantilever, it is fully fixed on one end. The length of the beam is 200 mm. All the dimensions are in mm.

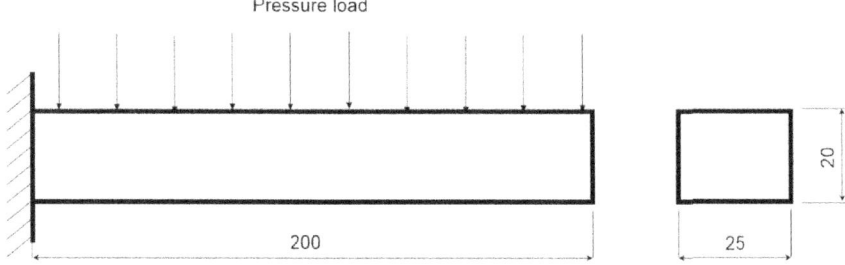

Figure 3.1 Cantilever beam under investigation

Let's start writing the Python script to do this task at hand.

3.2 Python script

In this section, the python script for finding the deflection of a cantilever beam with end load is presented. Comments have been added wherever necessary.

```
# Catilever beam bending under uniform pressure load

# The first three lines are required to import the required ABAQUS modules
# and create references to the objects that are defined by the module.
# The second line means that you are importing the symbolic constants
# (variables with a constant value) that have been defined by the scripting
# interface of Abaqus.It is a good practice to include these three lines at
# the top of every script that you write.The third line is used for acessing
# the objects of Region() method which is defined inside the regionToolset
# module.
```

```python
from abaqus import *
from abaqusConstants import*
import regionToolset

# This line is required to make the ABAQUS viewport display nothing.
# 'Viewport: 1' is the default name of an ABAQUS viewport. setValues()
# method is used to set the displayedObject parameter to None so that
# nothing is displayed.

session.viewports['Viewport: 1'].setValues(displayedObject=None)

# Model creation
# By default, ABAQUS creates a model named 'Model-1'. We will use the
# changeKey() method to change the name of the model. 'mdb' will give access
# to the model database and we will assign it ot 'cantileverModel'. The
# 'One  main word' strategy for naming variable will start from here.
# 'cantilever'will be the 'one main word' in this example,for naming main
# variables.

mdb.models.changeKey(fromName='Model-1', toName='Cantilever Beam')
cantileverModel = mdb.models['Cantilever Beam']

# Part creation

# These two statements will provide access to all the objects related to
# sketch and part. Including this is not mandatory, but is a good practice.

import sketch
import part

# In this set of statements, we would define a sketch by using the
# ConstrainedSketch() method. This method in turn has Rectangle() method
# which can be used to draw the rectangular cross section that we want in
# the sketch. Enter a value for 'sheetSize' based on your overall problem
# size.

cantileverSketch = cantileverModel.ConstrainedSketch(name='Beam Cross
Section', sheetSize=5)
```

```
cantileverSketch.rectangle(point1=(0.0,0.0), point2=(25.0,20.0))


# Using the sketch that we have created, let us create a 3D part using the
# Part() method. THREE_D is an example of symbolic constant.
# The BaseSolidExtrude() method is used to create a feature object based on
# the sketch we have created. The 'depth' value is the amount we extrude
# sketch. The extrusion, by default happens along the z - axis.

cantileverPart = cantileverModel.Part(name='Beam', dimensionality=THREE_D,
type=DEFORMABLE_BODY)
cantileverPart.BaseSolidExtrude(sketch=cantileverSketch, depth=200.0)


# Material definition

# First get access to objects relating to materials by using the import
# statement. Define the name of the material, which would be used further
# in the script. The Density() and Elastic() objects are used to specify
# density and elastic properties as name suggests. The input arguments to
# Density() looks so as it is actually a table with density values with
# respect to temperature. Here we don't need that. So we use blank spaces.
# The same logic applies to Elastic() where it is a table with Young's
# modulus values with respect to Poissons's ratio

import material
cantileverMaterial = cantileverModel.Material(name='Steel')
cantileverMaterial.Density(table=((7800, ),   ))
cantileverMaterial.Elastic(table=((200E9, 0.29),  ))

# Section creation and assignment
# Get access to the section objects by using the import statement. Then
# create a solid section by using the HomogenousSolidSection() method. You
# can see that we have refered to the material we created in the last step.

import section
cantileverSection =
cantileverModel.HomogeneousSolidSection(name='Cantilever Section',
material='Steel')
```

```python
# The next step is to assign the created section to cantilever members. For
# this we identify all the cells of the part and assign them to a region.
# The comma used here is to indicate that we are creating a Region object,
# which would be a sequence of cells. This sequence can be vertex objects,
# edge objects, node objetcs or face objects.

region_of_cantilever = (cantileverPart.cells,)

# Using the SectionAssignment() method, section assignment happens.

cantileverPart.SectionAssignment(region=region_of_cantilever,
sectionName='Cantilever Section')

# Assembly creation

# Get access to the assembly objects by using the import statement. The
# rootAssembly is an assembly object which is a member of the Model object.
# Create an instance of the part by using the Instance() method.By default,
# the 'dependent' parameter is set to OFF.Change this to ON.We have already
# defined the part name as 'cantileverPart'. We will refer to that now.

import assembly
cantileverAssembly = cantileverModel.rootAssembly
cantileverInstance = cantileverAssembly.Instance(name='Cantilever
Instance', part=cantileverPart, dependent=ON)

# Step creation

# Get access to the step objects by using the import statement. The
# StaticStep() method is used to create a static step which could be used
# for loading. This is the step next to 'Initial' step created by default.

import step
cantileverModel.StaticStep(name='Apply Pressure Load', previous='Initial',
description='Load is applied now')

# Definition of field output requests
```

```python
# ABAQUS, by default creates field output requests and names it as
# 'F-Output-1'. We will change it's name as well as set the parameters that
# we are interested in investigating.

cantileverModel.fieldOutputRequests.changeKey(fromName='F-Output-1',
toName='Required Field Outputs')
cantileverModel.fieldOutputRequests['Required Field
Outputs'].setValues(variables=('S', 'E', 'PEMAG', 'U', 'RF', 'CF'))

# History output request left at default

# Apply pressure loads
# First,We have to identify the face on which the pressure load needs to be
# applied. We will use the findAt() method to find the face by specifying
# the coordinate. Name the variables so that it is easy to understand.
point_on_topface_xcoord = 12.5
point_on_topface_ycoord = 20.0
point_on_topface_zcoord = 100.0

point_top_face =
(point_on_topface_xcoord,point_on_topface_ycoord,point_on_topface_zcoord)

# Refer Figure 3.1 to understand why we enter these values here. The
# coordinates are the midpoint of the top surface of the contilever beam
# geometry.

top_face = cantileverInstance.faces.findAt((point_top_face,))

# Now that we have identified the face, convert this into a region by using
# the regionToolset module.

top_face_region=regionToolset.Region(side1Faces=top_face)

# Using the newly created region, you can now apply the pressure load. You
# also have to specify the step name and magnitude of the pressure load.
```

```
cantileverModel.Pressure(name='Uniform Applied Pressure',
createStepName='Apply Pressure Load', region=top_face_region,
distributionType=UNIFORM, magnitude=0.5, amplitude=UNSET)


# Apply boundary conditions
# Similarly identify the face on the fixed end of the cantilever beam by
# its coordinates and create a region which could be used for defining the
# boundary conditions.

point_on_fixedface_xcoord = 12.5
point_on_fixedface_ycoord = 10.0
point_on_fixedface_zcoord = 0.0

point_on_fixed_face = (point_on_fixedface_xcoord,
point_on_fixedface_ycoord, point_on_fixedface_zcoord)
fixed_face = cantileverInstance.faces.findAt((point_on_fixed_face,))
fixed_face_region = regionToolset.Region(faces=fixed_face)
# Note that the boundary conditions are applied in the initial step.

cantileverModel.EncastreBC(name='Fix one end', createStepName='Initial',
region=fixed_face_region)


# Mesh creation

# Get access to the mesh objects by using the import statement. We will use
# the predefined regions for element type definition. C3D8R elements are
# used in this simulation.

import mesh
elem_type_for_mesh = mesh.ElemType(elemCode=C3D8R, elemLibrary=STANDARD,
kinematicSplit=AVERAGE_STRAIN, secondOrderAccuracy=OFF,
hourglassControl=DEFAULT, distortionControl=DEFAULT)

# The region for meshing is selected by selecting all the cells of the beam.
# This is, in turn, done by entering the coordinates of an internal point.

cantilever_interior_xcoord = 12.5
cantilever_interior_ycoord = 10.0
```

```python
cantilever_interior_zcoord = 100.0
cantileverCells = cantileverPart.cells
requiredcantileverCells =
cantileverCells.findAt((cantilever_interior_xcoord,cantilever_interior_ycoo
rd,cantilever_interior_zcoord),)


# As you can see, there is a comma at the end.This means that the input for
# the region is a sequence of cells. Once we have chosen the mesh region,
# the pre-defined element type is assigned which is followed by seeding the
# part. You can reduce the value of 'size' to generate a finer mesh.
# generateMesh() method is used to generate a mesh on the part.

cantilever_mesh_region=(requiredcantileverCells,)
cantileverPart.setElementType(regions=cantilever_mesh_region,
elemTypes=(elem_type_for_mesh,))
cantileverPart.seedPart(size=10.0, deviationFactor=0.1)
cantileverPart.generateMesh()

# Job creation and running
# Get access to the job objects by using the import statement. The Job()
# method is used to create a job. Make sure that you enter the correct name
# of the model. Most of the arguments entered here are not mandatory. You
# can edit the values base don your requirements.

import job
mdb.Job(name='CantilverJob', model='Cantilever Beam', type=ANALYSIS,
explicitPrecision=SINGLE, nodalOutputPrecision=SINGLE,
description='Simulating a cantilever beam',
parallelizationMethodExplicit=DOMAIN, multiprocessingMode=DEFAULT,
numDomains=1, userSubroutine='', numCpus=1, memory=50,
memoryUnits=PERCENTAGE, scratch='', echoPrint=OFF, modelPrint=OFF,
contactPrint=OFF, historyPrint=OFF)

# The submit() method is used for submitting the job for analysis. The
# waitForCompletion() makes ABAQUS wait till the job is fully executed.

mdb.jobs['CantilverJob'].submit(consistencyChecking=OFF)
mdb.jobs['CantilverJob'].waitForCompletion()
```

```
# Postprocessing

# Get access to the visualization objects by using the import statement.
# We save the odb object and path to variables which could be used for
# visualization. The node labels and element labels are turned on for
# better clarity.The viewport size can also be set.

import visualization
cantilever_viewport = session.Viewport(name='Cantilever Beam Results
Viewport')
cantilever_odb_path = 'CantilverJob.odb'
cantilever_odb_object = session.openOdb(name=cantilever_odb_path)
cantilever_viewport.setValues(displayedObject=cantilever_odb_object)
cantilever_viewport.odbDisplay.display.setValues(plotState=(DEFORMED, ))
```

3.3 Summary

The python script for finding the static response of the cantilever beam was presented with explanatory comments in this chapter. If you have some difficulties in understanding some of the things mentioned in the script, just note it down somewhere and keep moving forward. I am sure that, as you go further in this book, the doubts will get cleared though some other example.

But always keep in mind that writing easily readable Python scripts with properly named variables will make you more efficient in the long run. This will also help the people who will look into your work later.

Congratulations for having completed till this chapter. Hope you like the journey. Keep going!

4. Static / Dynamic analysis of a connecting lug

Things to learn

※ Making proper engineering assumptions to simplify the problem at hand

※ A technique to partition the cells

※ Writing a Python script to evaluate static deflection of a connecting lug

4.1 Problem description

This problem has been taken from the ABAQUS user manual. We will investigate the static response of the connecting lug which is presented in Figure 4.1. The lug is welded to a big structure on its left end. A bolt is expected to be placed on the 0.015m diameter hole. The objective is to find the static deflection of the lug when a 30 kN is applied to this bolt in the negative y direction.

As an engineering assumption, instead of actually modelling the bolt, we will use an uniform pressure load over the bottom half of the hole as shown in the figure to replicate the physical phenomenon. The variation of pressure around the hole circumference will also be neglected. The magnitude of the uniform pressure that will be applied will be 30 kN / (2 * 0.015m * 00.02m) = 50 MPa.

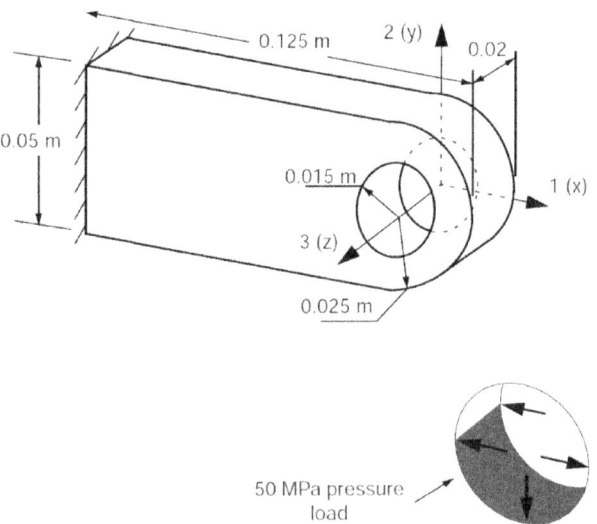

Figure 4.1 Dimensions of the connecting lug (DS Simulia, 2008)

The steps for doing this example using the GUI is explained in the ABAQUS manual.

4.2 Python script

```python
# Connecting lug with a uniform pressure load


# The first three lines are required to import the required ABAQUS modules
# and create references to the objects that are defined by the module.
# The second line means that you are importing the symbolic constants
# (variables with a constant value) that have been defined by the scripting
# interface of Abaqus. It is a good practice to include these three lines
# at the top of every script that you write. The third line is used for
# acessing the objects of Region() method which is defined inside the
# regionToolset module.

from abaqus import *
from abaqusConstants import*
import regionToolset

# This line is required to make the ABAQUS viewport display nothing.
# 'Viewport: 1' is the default name of an ABAQUS viewport. setValues()
# method is used to set the displayedObject parameter to None so that
# nothing is displayed.

session.viewports['Viewport: 1'].setValues(displayedObject=None)

# Model creation

# By default, ABAQUS creates a model named 'Model-1'. We will use the
# changeKey() method to change the name of the model.'mdb' will give access
# to the model database and we will assign it ot 'conLugModel'. The 'One
# main word' strategy for naming variable will start from here. 'conLug'
# will be the 'one main word' in this example, for naming main variables.

mdb.models.changeKey(fromName='Model-1', toName='Connecting Lug')
conLugModel = mdb.models['Connecting Lug']

# Part creation

# These two statements will provide access to all the objects related to
# sketch and part. Including this is not mandatory, but is a good practice.
```

```
import sketch
import part


# In this set of statements, we would define a sketch by using the
# ConstrainedSketch() method. This method in turn has Line() method
# which can be used to features that we want in the
# sketch. Enter a value for 'sheetSize' based on your overall problem size.

conLugProfileSketch = conLugModel.ConstrainedSketch(name='Conlug Profile',
sheetSize=1)
conLugProfileSketch.Line(point1=(0,0.025), point2=(-0.125,0.025))
conLugProfileSketch.Line(point1=(-0.125,0.025), point2=(-0.125,-0.025))
conLugProfileSketch.Line(point1=(-0.125,-0.025), point2=(0.0,-0.025))
conLugProfileSketch.ArcByCenterEnds(center=(0.0,0.0), point1=(0.0,-0.025),
point2=(0.0,0.025))
conLugProfileSketch.CircleByCenterPerimeter(center=(0.0, 0.0),
point1=(0.0,0.015))


# Using the sketch that we have created, let us create a 3D part using the
# Part() method. THREE_D is an example of symbolic constant.
# The BaseSolidExtrude() method is used to create a feature object based on
# the sketch we have created. The 'depth' value is the amount we extrude
# sketch. The extrusion, by default happens along the z - axis.

conLugPart = conLugModel.Part(name='Connecting lug',
dimensionality=THREE_D, type=DEFORMABLE_BODY)
conLugPart.BaseSolidExtrude(sketch=conLugProfileSketch, depth=0.02)


# Material definition
# First get access to objects relating to materials by using the import
# statement. Define the name of the material, which would be used further
# in the script. The Density() and Elastic() objects are used to specify
# density and elastic properties as name suggests. The input arguments to
# Density() looks so as it is actually a table with density values with
# respect to temperature. Here we don't need that. So we use blank spaces.
# The same logic applies to Elastic() where it is a table with Young's
# modulus values with respect to Poissons's ratio
```

```
import material
conLugMaterial = conLugModel.Material(name='Steel')
conLugMaterial.Density(table=((7800, ),      ))
conLugMaterial.Elastic(table=((200E9, 0.30), ))


# Section creation and assignment

# Get access to the section objects by using the import statement. Then
# create a solid section by using the HomogenousSolidSection() method. You
# can see that we have refered to the material we created in the last step.

import section
conLugSection = conLugModel.HomogeneousSolidSection(name='Con Lug Section',
material='Steel')

# The next step is to assign the created section to lug members. For
# this we identify all the cells of the part and assign them to a region.
# The comma used here is to indicate that we are creating a Region object,
# which would be a sequence of cells. This sequence can be vertex objects,
# edge objects, node objetcs or face objects.

conLug_region = (conLugPart.cells,)
conLugPart.SectionAssignment(region=conLug_region, sectionName='Con Lug
Section')

# Assembly creation

# Get access to the assembly objects by using the import statement. The
# rootAssembly is an assembly object which is a member of the Model object.
# Create an instance of the part by using the Instance() method. By default,
# the 'dependent' parameter is set to OFF. Change this to ON. We have
# already defined the part name as 'conLugPart'. We will refer to that now.

import assembly
conLugAssembly = conLugModel.rootAssembly
conLugInstance = conLugAssembly.Instance(name='conLug Instance',
part=conLugPart, dependent=ON)
```

```
# Step creation


# Get access to the step objects by using the import statement. The
# StaticStep() method is used to create a static step which could be used
# for loading. This is the step next to 'Initial' step created by default.

import step
conLugModel.StaticStep(name='Apply Load', previous='Initial',
description='Load is applied during this step')

# Definition of field output requests

# ABAQUS, by default creates field output requests and names it as
# 'F-Output-1'. We will change it's name as well as set the parameters that
# we are interested in investigating.

conLugModel.fieldOutputRequests.changeKey(fromName='F-Output-1',
toName='Required Field Outputs')
conLugModel.fieldOutputRequests['Required Field
Outputs'].setValues(variables=('S', 'E', 'PEMAG', 'U', 'RF', 'CF'))

# Definition of history output requests

# Create a new history output and delete the exsiting one. This is done
# only for illustration purposes.

conLugModel.HistoryOutputRequest(name='Default History Outputs',
createStepName='Apply Load', variables=PRESELECT)
del conLugModel.historyOutputRequests['H-Output-1']

# Identify the face by partitioning for load application

# In order to partition the part, we will create a datum plane and then use
# 'PartitionCellByDatumPlane'. Once we have partitioned, then we can use,
# the findAt() method to find the face and create a surface based on this.


conLugPart.DatumPlaneByPrincipalPlane(principalPlane=XZPLANE, offset=0.0)
```

```python
conLug_interior_xcoord = -0.0625
conLug_interior_ycoord = 0.0
conLug_interior_zcoord = 0.01
conLugCells = conLugPart.cells
allconLugCells =
conLugCells.findAt(((conLug_interior_xcoord,conLug_interior_ycoord,conLug_in
terior_zcoord),)
conLugPart.PartitionCellByDatumPlane(datumPlane=conLugPart.datums[3],
cells=allconLugCells)


# Finding the bottom curved surface so that it can be used for loading

conLug_bottomcurve_surface_point = (0.0,-0.015,0.01)
conLug_bottomcurve_surface =
conLugInstance.faces.findAt((conLug_bottomcurve_surface_point,))
conLugAssembly.Surface(side1Faces=conLug_bottomcurve_surface, name='Bottom
Curved Surface')
conLug_load_region = conLugAssembly.surfaces['Bottom Curved Surface']

# Apply the pressure loads.

conLugModel.Pressure(name='Load-1', createStepName='Apply Load',
region=conLug_load_region, distributionType=UNIFORM, magnitude=5.0E7,
amplitude=UNSET)

# Identify faces and apply boundary conditions

# Now that we have already partitioned the part along the XZ plane, we have
# to identify the two faces that represent the fixed end and fix them.

# Face 1
bc_face1_point_x = -0.125
bc_face1_point_y = 0.0125
bc_face1_point_z = 0.01

bc_face1_point = (bc_face1_point_x, bc_face1_point_y, bc_face1_point_z)
bc_face1 = conLugInstance.faces.findAt((bc_face1_point,))
```

```
bc_face1_region = regionToolset.Region(faces=bc_face1)


# Face 2

bc_face2_point_x = -0.125
bc_face2_point_y = -0.0125
bc_face2_point_z = 0.01


bc_face2_point = (bc_face2_point_x, bc_face2_point_y, bc_face2_point_z)
bc_face2 = conLugInstance.faces.findAt((bc_face2_point,))
bc_face2_region = regionToolset.Region(faces=bc_face2)

# Apply the encastre boundary conditions on both the faces to fix them.

conLugModel.EncastreBC(name='Encastre top face', createStepName='Initial',
region=bc_face1_region)
conLugModel.EncastreBC(name='Encastre bottom face',
createStepName='Initial', region=bc_face2_region)


# Mesh creation

# Get access to the mesh objects by using the import statement. We will use
# the predefined regions for element type definition. C3D8R elements are
# used in this simulation.

import mesh

elem_type_for_mesh = mesh.ElemType(elemCode=C3D20R, elemLibrary=STANDARD,
kinematicSplit=AVERAGE_STRAIN, secondOrderAccuracy=OFF,
hourglassControl=DEFAULT, distortionControl=DEFAULT)
conLugMeshRegion=(allconLugCells,)
conLugPart.setElementType(regions=conLugMeshRegion,
elemTypes=(elem_type_for_mesh,))
# You can reduce the value of 'size' to generate a finer mesh.
# generateMesh() method is used to generate a mesh on the part.

conLugPart.seedPart(size=0.005, deviationFactor=0.1)
conLugPart.generateMesh()
```

```python
# Job creation and running
# Get access to the job objects by using the import statement. The Job()
# method is used to create a job. Make sure that you enter the correct name
# of the model. Most of the arguments entered here are not mandatory. You
# can edit the values base don your requirements.

import job
mdb.Job(name='ConnectingLugJob', model='Connecting Lug', type=ANALYSIS,
explicitPrecision=SINGLE, nodalOutputPrecision=SINGLE,
description='Simulating a connecting lug',
parallelizationMethodExplicit=DOMAIN, multiprocessingMode=DEFAULT,
numDomains=1, userSubroutine='', numCpus=1, memory=50,
memoryUnits=PERCENTAGE, scratch='', echoPrint=OFF, modelPrint=OFF,
contactPrint=OFF, historyPrint=OFF)


# The submit() method is used for submitting the job for analysis. The
# waitForCompletion() makes ABAQUS wait till the job is fully executed.

mdb.jobs['ConnectingLugJob'].submit(consistencyChecking=OFF)
mdb.jobs['ConnectingLugJob'].waitForCompletion()


# Postprocessing
# Get access to the visualization objects by using the import statement.
# We save the odb object and path to variables which could be used for
# visualization. The node labels and element labels are turned on for
# better clarity.The viewport size can also be set.

import visualization
connecting_lug_viewport = session.Viewport(name='Connecting Lug Results
Viewport')
connecting_lug_Odb_Path = 'ConnectingLugJob.odb'
odb_object_1 = session.openOdb(name=connecting_lug_Odb_Path)
connecting_lug_viewport.setValues(displayedObject=odb_object_1)
connecting_lug_viewport.odbDisplay.display.setValues(plotState=(DEFORMED,
))
```

4.3 Summary

The partitioning strategy adopted in this problem was one of the fundamental ones and was done to facilitate structured meshing. This was one among many methods available. As an exercise, you can change the way the part is partitioned to have a much better mesh. Figure 4.2 was taken from ABAQUS Manual. You can try this out by making the required changes in the script and understand how the mesh changes with respect to your partitioning.

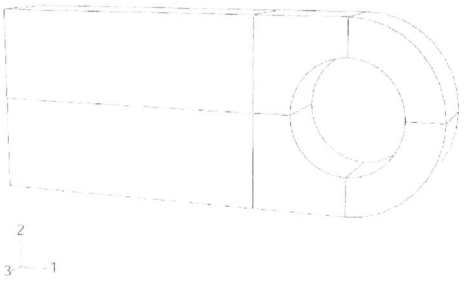

Figure 4.2 Try this partitioning by yourself

This brings us to the end of Chapter 4 – evaluating the static response of the connecting lug. Don't stop, keep moving forward!

5. Bending analysis of a plate

Things to learn

- ※ Mid-surface shell assumption
- ※ Partitioning the geometry
- ※ Writing a Python script to carry out a plate bending analysis

5.1 Problem description

In this chapter, let's write a Python script for simulating the bending of a plate of 0.1mm thickness. In this case, we will model the mid surface of the plate and discretize the geometry using shell elements.

The left end of the plate is fixed and the right edge is supported on roller joints. The dimensions can be seen from Figure 5.1. A pressure load is applied on this face as shown in the figure.

Here again, in this example, we would be partitioning the geometry and use the partitioned faces for definition and assignment. Let's get started. The material of the plate is generic steel.

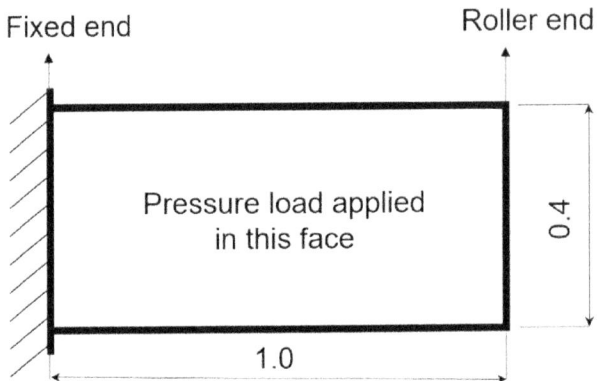

Figure 5.1 Simple plate geometry

5.2 Python script

```
# Simple plate bending - Shell
# The first three lines are required to import the required ABAQUS modules
# and create references to the objects that are defined by the module.
# The second line means that you are importing the symbolic constants
# (variables with a constant value) that have been defined by the scripting
# interface of Abaqus. It is a good practice to include these three lines
# at the top of every script that you write. The third line is used for
# acessing the objects of Region() method which is defined inside the
# regionToolset module.
```

```python
from abaqus import *
from abaqusConstants import*
import regionToolset


# This line is required to make the ABAQUS viewport display nothing.
# 'Viewport: 1' is the default name of an ABAQUS viewport. setValues()
# method is used to set the displayedObject parameter to None so that
# nothing is displayed.

# Model creation

# By default, ABAQUS creates a model named 'Model-1'. We will use the
# changeKey() method to change the name of the model. 'mdb' will give access
# to the model database and we will assign it ot 'simpleModel'. The 'One
# main word' strategy for naming variable will start from here. 'simple'
# will be the 'one main word' in this example, for naming main variables.

mdb.models.changeKey(fromName='Model-1', toName='Shell Bending Model')
simpleModel = mdb.models['Shell Bending Model']

# Part creation

# These two statements will provide access to all the objects related to
# sketch and part. Including this is not mandatory, but is a good practice.

import sketch
import part
# In this set of statements, we would define a sketch by using the
# ConstrainedSketch() method. This method in turn has Rectangle() method
# which can be used to draw the rectangle that we want in the
# sketch. Enter a value for 'sheetSize' based on your overall problem size.

simpleProfileSketch = simpleModel.ConstrainedSketch(name='Simple Plate
Sketch', sheetSize=3)
simpleProfileSketch.rectangle(point1=(0.0,0.0), point2=(1.0,0.4))

# Using the sketch that we have created, let us create a 3D part using the
# Part() method. THREE_D is an example of symbolic constant.
```

```
# The BaseShell() method is used to create a feature object based on
# the sketch we have created.

simplePart = simpleModel.Part(name='Plate', dimensionality=THREE_D,
type=DEFORMABLE_BODY)
simplePart.BaseShell(sketch=simpleProfileSketch)

# Material creation

# First get access to objects relating to materials by using the import
# statement. Define the name of the material, which would be used further
# in the script. The Density() and Elastic() objects are used to specify
# density and elastic properties as name suggests. The input arguments to
# Density() looks so as it is actually a table with density values with
# respect to temperature. Here we don't need that. So we use blank spaces.
# The same logic applies to Elastic() where it is a table with Young's
# modulus values with respect to Poissons's ratio

import material

simplePlateMaterial = simpleModel.Material(name='Steel')
simplePlateMaterial.Density(table=((7872, ),        ))
simplePlateMaterial.Elastic(table=((200E9, 0.29), ))

# Creation of a homogenous shell creation of 0.1 mmm thickness and it's
assignment

# Get access to the section objects by using the import statement. Then
# create a shell section by using the HomogenousShellSection() method. You
# can see that we have refered to the material we created in the last step.

import section
simplePlateSection = simpleModel.HomogeneousShellSection(name='Plate
Section', material ='Steel', thicknessType=UNIFORM, thickness=0.1)

# The next step is to assign the created section to lug members. For
# this we identify a point on the plate and assign them to a region.
# The comma used here is to indicate that we are creating a Region object,
```

```
# which would be a sequence of faces. This sequence can be vertex objects,
# edge objects, node objetcs or cell objects.

point_on_plate = (0.5, 0.2, 0.0)
face_on_plate = simplePart.faces.findAt((point_on_plate,))
region_of_plate = (face_on_plate,)

simplePart.SectionAssignment(region=region_of_plate, sectionName='Plate
Section', offset=0.0, offsetType=MIDDLE_SURFACE, offsetField='')

# Assembly creation

# Get access to the assembly objects by using the import statement. The
# rootAssembly is an assembly object which is a member of the Model object.
# Create an instance of the part by using the Instance() method. By default,
# the 'dependent' parameter is set to OFF. Change this to ON. We have
# already defined the part name as 'simplePart'. We will refer to that now.

import assembly

simpleAssembly = simpleModel.rootAssembly
simpleInstance = simpleAssembly.Instance(name='Plate Instance',
part=simplePart, dependent=ON)

# Step creation

# Get access to the step objects by using the import statement. The
# StaticStep() method is used to create a static step which could be used
# for loading. This is the step next to 'Initial' step created by default.

import step

simpleModel.StaticStep(name='Load Step', previous='Initial',
description='Apply pressure in this step', nlgeom=ON)

# Field output requests creation

# ABAQUS, by default creates field output requests and names it as
```

```
# 'F-Output-1'. We will set the parameters that
# we are interested in investigating.


simpleModel.fieldOutputRequests['F-Output-1'].setValues(variables=('S',
'RF', 'UT', 'U'))


# Partitioning
# Create the datum points by using DatumPointByCoordinate()


simplePart.DatumPointByCoordinate(coords=(0.5, 0.0, 0.0))
simplePart.DatumPointByCoordinate(coords=(0.5, 0.4, 0.0))


# Sort and store the created datum points accordingly


datums_keys = simplePart.datums.keys()
datums_keys.sort()
datum_point_1 = simplePart.datums[datums_keys[0]]
datum_point_2 = simplePart.datums[datums_keys[1]]


# Select the entire top face using findAt() and partition it using 2 points


point_on_face_to_partition = (0.5, 0.2, 0.0)
face_to_partition = simplePart.faces.findAt((point_on_face_to_partition,))
simplePart.PartitionFaceByShortestPath(point1=datum_point_1,
point2=datum_point_2, faces=face_to_partition)


# Identify the left edge and fix it using EncastreBC()


edge_to_fix = simpleInstance.edges.findAt(((0.0, 0.2, 0.0), ))
edge_to_fix_region = regionToolset.Region(edges=edge_to_fix)
simpleModel.EncastreBC(name='Encaster Edge', createStepName='Initial',
region=edge_to_fix_region)


# Identify the right edge and apply rolling condition using DisplacementBC()


edge_that_rolls = simpleInstance.edges.findAt(((1.0, 0.2, 0.0), ))
edge_that_rolls_region = regionToolset.Region(edges=edge_that_rolls)
```

```
simpleModel.DisplacementBC(name='Rolling Edge', createStepName='Initial',
region=edge_to_fix_region, u1=UNSET, u2=UNSET, ur3=SET, amplitude=UNSET,
distributionType=UNIFORM)

# Define faces using findAt() and then apply pressure on them

pressure_faces = simpleInstance.faces.findAt(((0.25, 0.2, 0.0), ), ((0.75,
0.2, 0.0), ))
pressure_faces_region = regionToolset.Region(side1Faces=pressure_faces)

simpleModel.Pressure(name='Apply Pressure', createStepName='Load Step',
region=pressure_faces_region, magnitude=2E3)

# Mesh creation

# Get access to the mesh objects by using the import statement. We will use
# the predefined regions for element type definition. S8R5 elements are
# used in this simulation.

import mesh

simple_mesh_region = region_of_plate
mesh_elem_type = mesh.ElemType(elemCode=S8R5, elemLibrary=STANDARD)
simplePart.setElementType(regions=simple_mesh_region,
elemTypes=(mesh_elem_type,))

# Identify the edges and then seeding them by number
# You can increase the value of 'number' to generate a finer mesh.

horizontal_edges = simplePart.edges.findAt(((0.25, 0.0, 0.0), ), ((0.75,
0.0, 0.0), ), ((0.75, 0.4, 0.0), ), ((0.25, 0.4, 0.0), ))
vertical_edges = simplePart.edges.findAt(((0.0, 0.2, 0.0), ), ((1.0, 0.2,
0.0), ))
simplePart.seedEdgeByNumber(edges=horizontal_edges, number=20)
simplePart.seedEdgeByNumber(edges=vertical_edges, number=16)

# generateMesh() method is used to generate a mesh on the part.
simplePart.generateMesh()
```

```python
# Job creation and running
# Get access to the job objects by using the import statement. The Job()
# method is used to create a job. Make sure that you enter the correct name
# of the model.

import job

mdb.Job(name='SimplePlateJob', model='Shell Bending Model', type=ANALYSIS,
description='Job simulates the bending of a plate under a load of 2MPa')

# The submit() method is used for submitting the job for analysis. The
# waitForCompletion() makes ABAQUS wait till the job is fully executed.

mdb.jobs['SimplePlateJob'].submit(consistencyChecking=OFF)
mdb.jobs['SimplePlateJob'].waitForCompletion()

# Post processing

# Get access to the visualization objects by using the import statement.
# We save the odb object and path to variables which could be used for
# visualization. The node labels and element labels are turned on for
# better clarity.The viewport size can also be set.

import visualization

simple_plate_viewport = session.Viewport(name='Plate Bending Results Viewport')
simple_plate_Path = 'SimplePlateJob.odb'
simple_odb_object = session.openOdb(name=simple_plate_Path)
simple_plate_viewport.setValues(displayedObject=simple_odb_object)
simple_plate_viewport.odbDisplay.display.setValues(plotState=(CONTOURS_ON_DEF, ))
```

5.3 Summary

A simple plate, subjected to uniform pressure load on its top face was simulated by modelling the mid surface and discretizing it using shell elements. The sort() method was used in this example.

If you tried out the 4 scripts that have been discussed so far, I am sure that you would have gotten a good feel of writing Python scripts for ABAQUS. You can observe that some of the lines are similar in every example. This is exactly the advantage of carrying out the simulations using Python scripting. You can automate a lot of repetitive actions and become more productive.

You are awesome, keep moving forward!

6. Contact analysis of an electrical switch

Things to learn

- ※ Making proper engineering assumptions
- ※ Defining the element normals, master and slave surfaces
- ※ Writing a Python script to carry out a contact analysis

6.1 Problem description

In this chapter, let's write a Python script for simulating the contact analysis of an electrical switch shown in Figure 6.1. The upper portion of the switch is displaced by a defined amount which brings it in contact with the lower part. The contacting region will be the subject matter of investigation. The plates are made of generic steel. An engineering assumption is made in this problem: The plate on the right side of the structure (perpendicular to two plates) will not be modelled as it is fixed and has no influence on the predictions. The dimensions are in mm.

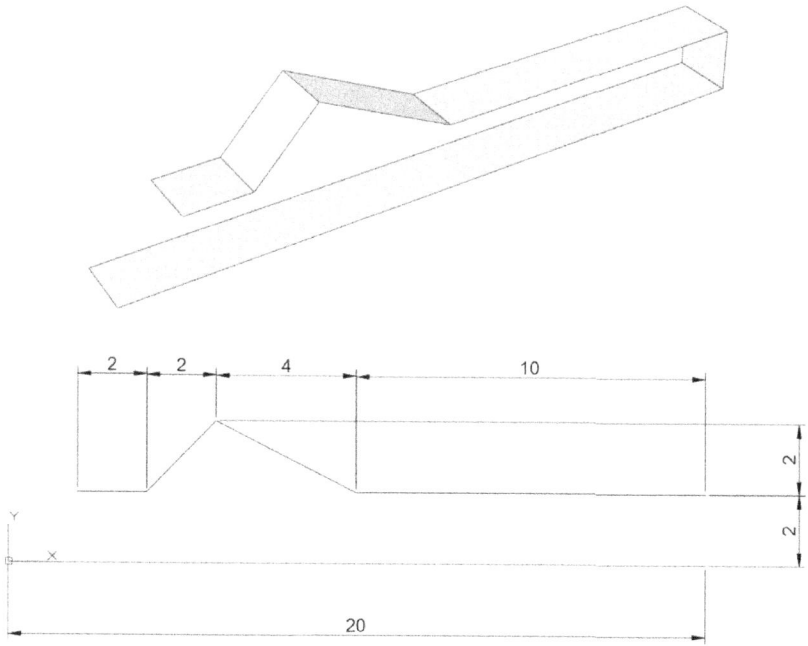

Figure 6.1 Schematic of an electrical switch

Note:

This example was taken from
https://www.engr.mun.ca/~adluri/courses/fem/7706/Abaqus/Tutorial%206.pdf

The step by step procedure for doing the analysis in the GUI is explained in the above link.

6.2 Python script

```
# Contact analysis of an electrical switch


# The first three lines are required to import the required ABAQUS modules
# and create references to the objects that are defined by the module.
# The second line means that you are importing the symbolic constants
# (variables with a constant value) that have been defined by the scripting
# interface of Abaqus. It is a good practice to include these three lines
# at the top of every script that you write. The third line is used for
# acessing the objects of Region() method which is defined inside the
# regionToolset module.

from abaqus import*
from abaqusConstants import*
import regionToolset

session.viewports['Viewport: 1'].setValues(displayedObject=None)

# This line is required to make the ABAQUS viewport display nothing.
# 'Viewport: 1' is the default name of an ABAQUS viewport. setValues()
# method is used to set the displayedObject parameter to None so that
# nothing is displayed.

# Model creation

# By default, ABAQUS creates a model named 'Model-1'. We will use the
# changeKey() method to change the name of the model. 'mdb' will giveaccess
# to the model database and we will assign it ot 'switchModel'. The 'One
# main word' strategy for naming variable will start from here. 'switch'
# will be the 'one main word' in this example, for naming main variables.

mdb.models.changeKey(fromName='Model-1', toName='Electrical Switch Model')
switchModel = mdb.models['Electrical Switch Model']

# Part creation

# These two statements will provide access to all the objects related to
# sketch and part. Including this is not mandatory, but is a good practice.
```

```python
import sketch
import part


# In this set of statements, we would define a sketch by using the
# ConstrainedSketch() method. This method in turn has Line() method
# which can be used to draw the lines that we want in the
# sketch. Enter a value for 'sheetSize' based on your overall problem size.

switchProfileSketch = switchModel.ConstrainedSketch(name='Electrical Switch
Sketch', sheetSize=50)
switchProfileSketch.Line(point1=(0.0,0.0), point2=(-20.0,0.0))
switchProfileSketch.Line(point1=(0.0,2.0), point2=(-10.0,2.0))
switchProfileSketch.Line(point1=(-10.0,2.0), point2=(-14.0,4.0))
switchProfileSketch.Line(point1=(-14.0,4.0), point2=(-16.0,2.0))
switchProfileSketch.Line(point1=(-16.0,2.0), point2=(-18.0,2.0))

# Using the sketch that we have created, let us create a 3D part using the
# Part() method. THREE_D is an example of symbolic constant.
# The BaseShellExtrude() method is used to create a feature object based on
# the sketch we have created with a depth of 2

switchPart = switchModel.Part(name='Switch Part', dimensionality=THREE_D,
type=DEFORMABLE_BODY)
switchPart.BaseShellExtrude(sketch=switchProfileSketch, depth=2)

# Material creation

# First get access to objects relating to materials by using the import
# statement. Define the name of the material, which would be used further
# in the script. The Density() and Elastic() objects are used to specify
# density and elastic properties as name suggests. The input arguments to
# Density() looks so as it is actually a table with density values with
# respect to temperature. Here we don't need that. So we use blank spaces.
# The same logic applies to Elastic() where it is a table with Young's
# modulus values with respect to Poissons's ratio
```

```
import material

switchMaterial = switchModel.Material(name='Generic Steel')
switchMaterial.Elastic(table=((210E9, 0.30), ))


# Creation of a homogenous shell creation of 0.15 mmm thickness and it's
# assignment

# Get access to the section objects by using the import statement. Then
# create a shell section by using the HomogenousShellSection() method. You
# can see that we have refered to the material we created in the last step.

import section

switchSection = switchModel.HomogeneousShellSection(name='Switch Section',
material ='Generic Steel', thicknessType=UNIFORM, thickness=0.15)

# Identify all the faces by using the findAt() method

point1 = (-10.0,0.0,0.0)
point2 = (-5.0,2.0,0.0)
point3 = (-12.0,3.0,0.0)
point4 = (-15.0,3.0,0.0)
point5 = (-17.0,2.0,0.0)
all_faces = switchPart.faces.findAt((point1,), (point2,), (point3,),
(point4,), (point5,))
switchRegion = (all_faces,)

switchPart.SectionAssignment(region=switchRegion, sectionName='Switch
Section', offset=0.0, offsetType=MIDDLE_SURFACE, offsetField='')

# For contact analysis, you will have to define the element normals. This
# should be done by using the flipNormal() method. Finding the direction of
# the face orientation can be challenging. So it is recommended to try upto
# this point, find if the flipping direction is correct or not and then
# proceed further. We have to make sure that purple colored faces (can be
# seen in the GUI) must be facing each other. The regionToolset module is
# used to define the region and then assign the element normals.
```

```
flippingface_point = (-10.0,0.0,0.0)
flippingface = switchPart.faces.findAt((flippingface_point,))
flippingface_region = regionToolset.Region(faces=flippingface)
switchPart.flipNormal(regions=flippingface_region)

# Assembly creation

# Get access to the assembly objects by using the import statement. The
# rootAssembly is an assembly object which is a member of the Model object.
# Create an instance of the part by using the Instance() method. By default,
# the 'dependent' parameter is set to OFF. Change this to ON. We have
# already defined the part name as 'switchPart'. We will refer to that now.

import assembly

switchAssembly = switchModel.rootAssembly
switchInstance = switchAssembly.Instance(name='Switch Instance',
part=switchPart, dependent=ON)

# Step creation

# Get access to the step objects by using the import statement. The
# StaticStep() method is used to create a static step which could be used
# for loading. This is the step next to 'Initial' step created by default.

import step

switchModel.StaticStep(name='Load Step', previous='Initial',
description='Applying forces in this step', nlgeom=ON)

# Application of boundary conditions

# Identify the two edges on the right side, define a region using the
# regionToolset module and use a EncastreBC() method to fix the edges

rightedge_point1 = (0.0,0.0,1.0)
rightedge_point2 = (0.0,2.0,1.0)
```

```
rightedges = switchInstance.edges.findAt((rightedge_point1,),
(rightedge_point2,))
rightedges_region = regionToolset.Region(edges=rightedges)
switchModel.EncastreBC(name='Encaster Edge', createStepName='Initial',
region=rightedges_region)

# Next,we have to apply displacement boundary condition to an edge.For this
# we will identify the edge and create a Set or region based on the edge.
# This region can be used for applying the boundary condition.

displacementedge_point = (-14.0,4.0,1.0)
displacementedge = switchInstance.edges.findAt((displacementedge_point,))
displacementedge_region = switchAssembly.Set(edges=displacementedge,
name='Displacement Set')
switchModel.DisplacementBC(name='Displacement BC', createStepName='Load
Step', region=displacementedge_region, u1=UNSET, u2=-3.0, u3=UNSET,
ur1=UNSET, ur2=UNSET, ur3=UNSET, amplitude=UNSET, distributionType=UNIFORM,
fieldName='', localCsys=None)

# Datum plane creation and partitioning. The bottom face will be
# partitioned and these two faces will be used for mesh and contact
# definition. Note that we are using the already defined 'point1' here.

switchPart.DatumPlaneByPrincipalPlane(principalPlane=YZPLANE, offset=-10.0)
face_to_partition = switchPart.faces.findAt((point1,))
switchPart.PartitionFaceByDatumPlane(datumPlane=switchPart.datums[4],
faces=face_to_partition)

# Defining Interaction properties

# The contact properties like Tangential and normal behavior have to be
# defined.A frictionless tangential behavior is assumed. Default parameters
# are used for defining the Normal behavior. Refer the Abaqus scripting
# manual for syntax and required input arguments.

import interaction

switchModel.ContactProperty('Interaction Property')
```

```python
switchModel.interactionProperties['Interaction
Property'].TangentialBehavior(formulation=FRICTIONLESS)
switchModel.interactionProperties['Interaction
Property'].NormalBehavior(pressureOverclosure=HARD, allowSeparation=ON,
contactStiffness=DEFAULT, contactStiffnessScaleFactor=1.0,
clearanceAtZeroContactPressure=0.0, stiffnessBehavior=LINEAR,
constraintEnforcementMethod=PENALTY)


# Two points, each for defining the master and slave surfaces are selected.
# Using these two points, the faces are identified and therefore the
# surfaces are defined respectively.

master_surface_point = (-15.0,0.0,1.0)
master_surface = switchInstance.faces.findAt((master_surface_point,))
master_surface_region = switchAssembly.Surface(side1Faces=master_surface,
name='Master Surface')

slave_surface_point = (-17.0,2.0,1.0)
slave_surface = switchInstance.faces.findAt((slave_surface_point,))
slave_surface_region = switchAssembly.Surface(side1Faces=slave_surface,
name='Slave Surface')

# Refer the scripting manual for the correct usage of this statement.

switchModel.SurfaceToSurfaceContactStd(name='SurfacetoSurfaceContact',
createStepName='Initial', master=master_surface_region,
slave=slave_surface_region, sliding=FINITE, thickness=ON,
interactionProperty='Interaction Property', adjustMethod=NONE,
initialClearance=OMIT, datumAxis=None, clearanceRegion=None)

# Seeding edges and meshing

# Get access to the mesh objects by using the import statement. We will use
# the predefined regions for element type definition. S4R elements are used
# in this simulation.

import mesh
```

```python
elemType1 = mesh.ElemType(elemCode=S4R, elemLibrary=STANDARD,
secondOrderAccuracy=OFF, hourglassControl=DEFAULT)
pickedRegions = switchPart.faces.getSequenceFromMask(mask=('[#3f ]', ), )
switchPart.setMeshControls(regions=pickedRegions, elemShape=QUAD,
technique=STRUCTURED)

# Seeding all edges. Reduce the value of 'size' to obtain finer mesh.
# generateMesh() method is used to generate a mesh on the part.

switchPart.seedEdgeBySize(edges=switchInstance.edges, size=0.25,
deviationFactor=0.1, constraint=FINER)
switchPart.generateMesh()

# Job creation
# Get access to the job objects by using the import statement. The Job()
# method is used to create a job. Make sure that you enter the correct name
# of the model.

import job

mdb.Job(name='switchContactJob', model='Electrical Switch Model',
description='Contact analysis of a switch', type=ANALYSIS, memory=90,
memoryUnits=PERCENTAGE, getMemoryFromAnalysis=True,
explicitPrecision=SINGLE, nodalOutputPrecision=SINGLE, echoPrint=OFF,
modelPrint=OFF, contactPrint=OFF, historyPrint=OFF, userSubroutine='',
scratch='', resultsFormat=ODB, multiprocessingMode=DEFAULT, numCpus=1,
numGPUs=0)

# The submit() method is used for submitting the job for analysis. The
# waitForCompletion() makes ABAQUS wait till the job is fully executed.

mdb.jobs['switchContactJob'].submit(consistencyChecking=OFF)
mdb.jobs['switchContactJob'].waitForCompletion()

# Post processing

# Get access to the visualization objects by using the import statement.
# We save the odb object and path to variables which could be used for
```

```
# visualization. The node labels and element labels are turned on for
# better clarity.The viewport size can also be set.

import visualization

switchViewport = session.Viewport(name='Switch contact analysis Viewport')
switch_Odb_Path = 'switchContactJob.odb'
odb_object = session.openOdb(name=switch_Odb_Path)
switchViewport.setValues(displayedObject=odb_object)
switchViewport.odbDisplay.display.setValues(plotState=(CONTOURS_ON_DEF, ))
```

6.3 Summary

A python script for carrying out the contact analysis of an electrical switch was presented with detailed explanatory comments. A displacement boundary condition was applied on one of the edges and two edges were totally fixed. The technique of selecting the master and slave surfaces through python scripting was explained in this example.

Great going so far, rock on!

7. Wireframe analysis(3D) with box profile

Things to learn

※ Create two parts and join them using connector elements

※ Using the sort() method to sort the datum objects

※ Creating and assigning section profiles

※ Using a for loop to create tuples

※ Writing a Python script to carry out a 3D wireframe analysis with box profile

7.1 Problem description

In this chapter, let's write a Python script for simulating a 3D wireframe model with a box profile under a static load. The problem statement is slightly modified version of a 3D wireframe tutorial available at this website(https://www.simuleon.com/abaqus-tutorials/).

A load of 100 N is applied on the four vertices depicted by red arrows in Figure 7.1. The blue dots in the figure represent the fixed vertices. The material of the wireframe is chosen as steel. In this analysis, we will model two parts. The first part will consist of the two 200 X 100 rectangular frames. The second part will have the 4 cross linking members that connect the two rectangular frames of the first part. We will join them using connectors, apply the load, boundary conditions and then solve the problem by writing a Python script.

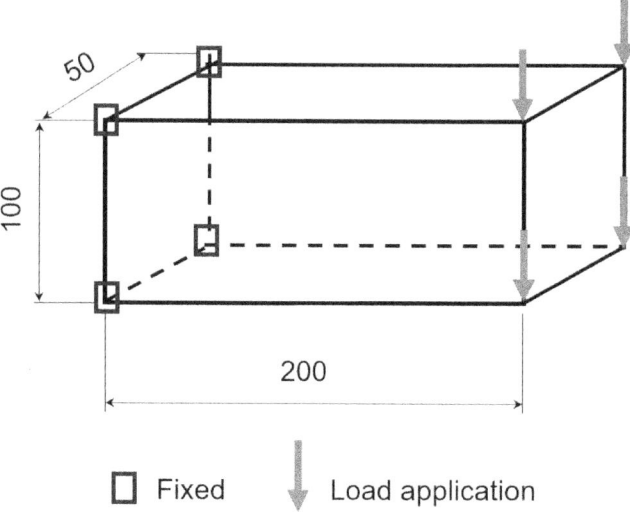

Figure 7.1 Schematic of the 3D wireframe model

A box profile for a section is defined with parameters such as width(a), height(b) and thickness. The details of these values for the two profiles used in this example is given in Table 7.1.

Part	Width(a)	Height(b)	Thickness(mm)
Rectangular frame	10.0	10.0	1.0
Cross links	5.0	5.0	1.0

Table 7.1 Box profile parameters used in this example

7.2 Python script

```
# Wireframe analysis

# The first three lines are required to import the required ABAQUS modules
# and create references to the objects that are defined by the module.
# The second line means that you are importing the symbolic constants
# (variables with a constant value) that have been defined by the scripting
# interface of Abaqus. It is a good practice to include these three lines
# at the top of every script that you write. The third line is used for
# acessing the objects of Region() method which is defined inside the
# regionToolset module.

from abaqus import*
from abaqusConstants import*
import regionToolset

session.viewports['Viewport: 1'].setValues(displayedObject=None)

# This line is required to make the ABAQUS viewport display nothing.
# 'Viewport: 1' is the default name of an ABAQUS viewport. setValues()
# method is used to set the displayedObject parameter to None so that
# nothing is displayed.

# Model creation

# By default, ABAQUS creates a model named 'Model-1'. We will use the
# changeKey() method to change the name of the model. 'mdb' will giveaccess
# to the model database and we will assign it ot 'wireModel'. The 'One
```

```python
# main word' strategy for naming variable will start from here. 'wire'
# will be the 'one main word' in this example, for naming main variables.

mdb.models.changeKey(fromName='Model-1', toName='Frame analysis')
wireModel = mdb.models['Frame analysis']

# Part creation

# These two statements will provide access to all the objects related to
# sketch and part. Including this is not mandatory, but is a good practice.

import sketch
import part

# Create the Frame using the Part() method. First define a reference point
# and using this reference point, we can create some datum points

wirePart = wireModel.Part(name='Frame', dimensionality=THREE_D,
type=DEFORMABLE_BODY)
wirePart.ReferencePoint(point=(0.0, 0.0, 0.0))

# Creating one side of the frame
# Create other datum points by offsetting from the reference point using a
# vector.

first_reference_point = wirePart.referencePoints[1]
wirePart.DatumPointByOffset(point= first _reference_point, vector=(0.0,
0.0, 0.0))
wirePart.DatumPointByOffset(point= first _reference_point, vector=(200.0,
0.0, 0.0))
wirePart.DatumPointByOffset(point= first _reference_point, vector=(200.0,
100.0, 0.0))

# The above created keys which has information about the newly created
# datum points might be in random order.
# Sort them in ascending order. After we get the points, a datum plane can
# be created using the 3 points.
```

```python
wirePart_datumKeys = wirePart.datums.keys()
wirePart_datumKeys.sort()
frame_datum_point_1 = wirePart.datums[wirePart_datumKeys[2]]
frame_datum_point_2 = wirePart.datums[wirePart_datumKeys[1]]
frame_datum_point_3 = wirePart.datums[wirePart_datumKeys[0]]
wirePart.DatumPlaneByThreePoints(point1=frame_datum_point_1, point2=
frame_datum_point_2, point3= frame_datum_point_3)


# Create a datum axis
wirePart.DatumAxisByPrincipalAxis(principalAxis=YAXIS)


# Now there will be totally 5 objects in the datums repository
# 3 datum points, 1 datum plane & 1 datum axis
# The key of Datum axis will be the highest number
# The key of Datum plane will be the second highest number which will be
# followed by keys of Datum points. You can get the keys by using the
# keys() method. Once you get the keys, sort them using the sort() method.

wirePart_datumKeys = wirePart.datums.keys()
wirePart_datumKeys.sort()


# Using the sorted keys and applying the logic mentioned in the comments
# above, we can get the index of datum plane and datum axis. Note that
# we are using the len() statement. Also remember that list indexing in
# Python starts from zero. So we get the axis index by subtracting one
# from the length of the list and so on..

index_of_plane = (len(wirePart_datumKeys) - 2)
index_of_axis = (len(wirePart_datumKeys) - 1)
frame_datum_plane = wirePart.datums[wirePart_datumKeys[index_of_plane]]
frame_datum_axis = wirePart.datums[wirePart_datumKeys[index_of_axis]]


# We now have the datum plane and datum axis ready, on which we
# can create the sketch using the MakeSketchTransform() method.

sketch_transform1 =
wirePart.MakeSketchTransform(sketchPlane=frame_datum_plane,
```

```python
              sketchUpEdge=frame_datum_axis, sketchPlaneSide=SIDE2,
              sketchOrientation=RIGHT, origin=(0.0, 0.0, 0.0))
wirePart_sketch = wireModel.ConstrainedSketch(name='frame sketch 1',
              sheetSize=250, gridSpacing=1, transform=sketch_transform1)

# Drawing the rectangle

wirePart_sketch.rectangle(point1=(0.0,0.0), point2=(200.0,100.0))

# Once you completed the sketch, use it to create a wire as below

wirePart.Wire(sketchPlane=frame_datum_plane, sketchUpEdge=frame_datum_axis,
              sketchPlaneSide=SIDE2, sketchOrientation=RIGHT, sketch=wirePart_sketch)

# Scripting for the second rectangular frame starts here...
# Apply the same logic explained above and create the other side of the
# frame. As the same procedure is carried out again, comments are avoided
# here.

wirePart.DatumPlaneByOffset(plane=frame_datum_plane, flip=SIDE2,
              offset=50.0)
wirePart_datumKeys = wirePart.datum.keys()
wirePart_datumKeys.sort()
index_of_plane2 = (len(wirePart_datumKeys) - 1)
frame_datum_plane2 = wirePart.datums[wirePart_datumKeys[index_of_plane2]]

wirePart.DatumPointByCoordinate(coords=(0.0, 0.0, 50.0))
wirePart.DatumPointByCoordinate(coords=(200.0, 0.0, 50.0))
wirePart.DatumPointByCoordinate(coords=(200.0, 100.0, 50.0))

wirePart.DatumAxisByTwoPoint(point1=(0.0,0.0,50.0),
              point2=(0.0,100.0,50.0))

wirePart_datumKeys = wirePart.datums.keys()
wirePart_datumKeys.sort()
index_of_axis2 = (len(wirePart_datumKeys) -1)
frame_datum_axis2 = wirePart.datums[wirePart_datumKeys[index_of_axis2]]
```

```
# Creation of sketch
sketch_transform2 =
wirePart.MakeSketchTransform(sketchPlane=frame_datum_plane2,
sketchUpEdge=frame_datum_axis2, sketchPlaneSide=SIDE1,
sketchOrientation=LEFT, origin=(0.0, 0.0, 50.0))
wirePart_sketch2 = wireModel.ConstrainedSketch(name='frame sketch 2',
sheetSize=20, gridSpacing=1, transform=sketch_transform2)

# Drawing the lines
wirePart_sketch2.rectangle(point1=(0.0,0.0), point2=(200.0,100.0))

# Use the sketch to create a wire
wirePart.Wire(sketchPlane=frame_datum_plane2,
sketchUpEdge=frame_datum_axis2, sketchPlaneSide=SIDE1,
sketchOrientation=LEFT, sketch=wirePart_sketch2)

# Now that we have created the two rectangular frames, we can create the
# cross links that connect the two frames as a separate part.

linkPart = wireModel.Part(name='Crosslinks', dimensionality=THREE_D,
type=DEFORMABLE_BODY)

linkPart.DatumPointByCoordinate(coords=(0.0, 0.0, 0.0))
linkPart.DatumPointByCoordinate(coords=(200.0, 0.0, 0.0))
linkPart.DatumPointByCoordinate(coords=(200.0, 100.0, 0.0))
linkPart.DatumPointByCoordinate(coords=(0.0, 100.0, 0.0))
linkPart.DatumPointByCoordinate(coords=(0.0, 0.0, 50.0))
linkPart.DatumPointByCoordinate(coords=(200.0, 0.0, 50.0))
linkPart.DatumPointByCoordinate(coords=(200.0, 100.0, 50.0))
linkPart.DatumPointByCoordinate(coords=(0.0, 100.0, 50.0))

# Sort the datum points that have been created and use them to create
# a wirePolyLine as shown below. In this way, we will create the second
# part

linkPart_datumKeys = linkPart.datums.keys()
linkPart_datumKeys.sort()
datum_points = linkPart.datums
```

```python
linkPart.WirePolyLine(points=((datum_points[linkPart_datumKeys[0]],
datum_points[linkPart_datumKeys[4]]), (datum_points[linkPart_datumKeys[1]],
datum_points[linkPart_datumKeys[5]]), (datum_points[linkPart_datumKeys[2]],
datum_points[linkPart_datumKeys[6]]), (datum_points[linkPart_datumKeys[3]],
datum_points[linkPart_datumKeys[7]])), mergeType=IMPRINT, meshable=ON)


# Material creation


# First get access to objects relating to materials by using the import
# statement. Define the name of the material, which would be used further
# in the script. The Density() and Elastic() objects are used to specify
# density and elastic properties as name suggests. The input arguments to
# Density() looks so as it is actually a table with density values with
# respect to temperature. Here we don't need that. So we use blank spaces.
# The same logic applies to Elastic() where it is a table with Young's
# modulus values with respect to Poissons's ratio

import material

wireMaterial = wireModel.Material(name='Steel')
wireMaterial.Elastic(table=((200E9, 0.29), ))


# Profile creation
# The box profiles are created based on the parameters defined in Table
# 7.1.

wireModel.BoxProfile(name='MainProfile', a=10.0, b=10.0,
uniformThickness=ON, t1=1.0)
wireModel.BoxProfile(name='LinkProfile', a=5.0, b=5.0, uniformThickness=ON,
t1=1.0)
# Sections creation and assignment

# Get access to the section objects by using the import statement. Then
# create a beam section by using the BeamSection() method. You
# can see that we have refered to the material and the profile we created
# in the last step.

import section
```

```
wireModel.BeamSection(name='Main Section', integration=DURING_ANALYSIS,
poissonRatio=0.0, profile='MainProfile', material='Steel',
temperatureVar=LINEAR, consistentMassMatrix=False)
wireModel.BeamSection(name='Link Section', integration=DURING_ANALYSIS,
poissonRatio=0.0, profile='LinkProfile', material='Steel',
temperatureVar=LINEAR, consistentMassMatrix=False)


# We have to identify the edges of the section so that we can assign them
# newly created sections. The findAt() method is used to find the edges,
# then we create a region which is used for section assignment.

mainSection_edges = wirePart.edges.findAt(((0.0, 50.0, 0.0), ), ((0.0,
50.0, 50.0), ), ((100.0, 0.0, 0.0), ), ((100.0, 0.0, 50.0), ), ((200.0,
50.0, 0.0), ), ((200.0, 50.0, 50.0), ), ((100.0, 100.0, 0.0), ), ((100.0,
100.0, 50.0), ),)
main_region = wirePart.Set(edges=mainSection_edges, name='Main Set')
wirePart.SectionAssignment(region=main_region, sectionName='Main Section',
offset=0.0, offsetType=MIDDLE_SURFACE, offsetField='',
thicknessAssignment=FROM_SECTION)


# The same procedure is done for the second part as well - linkPart

linkSection_edges = linkPart.edges.findAt(((0.0, 0.0, 25.0), ), ((0.0,
100.0, 25.0), ), ((200.0, 0.0, 25.0), ), ((200.0, 100.0, 25.0), ),)
link_region = linkPart.Set(edges=linkSection_edges, name='Link Set')
linkPart.SectionAssignment(region=link_region, sectionName='Link Section',
offset=0.0, offsetType=MIDDLE_SURFACE, offsetField='',
thicknessAssignment=FROM_SECTION)



# The beam orientation has to be defined for both of the regions from the
# two parts.

wirePart.assignBeamSectionOrientation(region=main_region,
method=N1_COSINES, n1=(0.0, 0.0, 1.0))
linkPart.assignBeamSectionOrientation(region=link_region,
method=N1_COSINES, n1=(1.0, 0.0, 0.0))
```

```
# Assembly creation


# Get access to the assembly objects by using the import statement. The
# rootAssembly is an assembly object which is a member of the Model object.
# Create an instance of the part by using the Instance() method.By default,
# the 'dependent' parameter is set to OFF. Change this to ON. We have
# already defined the part name as 'wirePart'. We will refer to that now.
# And do the same for linkPart also.

import assembly

wireAssembly = wireModel.rootAssembly
wireInstance = wireAssembly.Instance(name='Main Instance', part=wirePart,
dependent=ON)
linkInstance = wireAssembly.Instance(name='Link Instance', part=linkPart,
dependent=ON)


# Creation of wire features
# In this section, we will have to define the connections bewteen the two
# parts. For this, we identify the repsective vertices from both the parts
# using the findAt() method as done below.

import interaction

vertices_wirePart_side1 = wireInstance.vertices.findAt(((0.0, 0.0, 0.0),),
((200.0, 0.0, 0.0),), ((200.0, 100.0, 0.0),), ((0.0, 100.0, 0.0),),)
vertices_wirePart_side2 = wireInstance.vertices.findAt(((0.0, 0.0, 50.0),),
((200.0, 0.0, 50.0),), ((200.0, 100.0, 50.0),), ((0.0, 100.0, 50.0),),)
vertices_linkPart_side1 = linkInstance.vertices.findAt(((0.0, 0.0, 0.0),),
((200.0, 0.0, 0.0),), ((200.0, 100.0, 0.0),), ((0.0, 100.0, 0.0),),)
vertices_linkPart_side2 = linkInstance.vertices.findAt(((0.0, 0.0, 50.0),),
((200.0, 0.0, 50.0),), ((200.0, 100.0, 50.0),), ((0.0, 100.0, 50.0),),)

# Now create the connectors using the above points.For creating connectors,
# we need to have a tuple of all the vertices. So for this, we will use a
# for loop and append all the vertices into a variable 'vertices_list'.

vertices_list = []
```

```
for i in range(len(vertices_wirePart_side1)):
    vertices_list.append((vertices_wirePart_side1[i],
vertices_linkPart_side1[i]))
    vertices_list.append((vertices_wirePart_side2[i],
vertices_linkPart_side2[i]))

# Once we have the list of vertices,we can create a tuple using the tuple()
# and then use the WirePolyLine() method to create the wire poly lines.

pointtuples = tuple(vertices_list)
wireAssembly.WirePolyLine(points=pointtuples, mergeType=IMPRINT,
meshable=OFF)

# We need all the connector edges so that we can define a set and use it
# definition of the joints as shown below.

connector_edges = wireAssembly.edges.findAt(((0.0, 0.0, 0.0),), ((200.0,
0.0, 0.0),), ((200.0, 100.0, 0.0),), ((0.0, 100.0, 0.0),), ((0.0, 0.0,
50.0),), ((200.0, 0.0, 50.0),), ((200.0, 100.0, 50.0),), ((0.0, 100.0,
50.0),),)
wireAssembly.Set(edges=connector_edges, name='Connector wires Set')

# Creation of a connector section

wireModel.ConnectorSection(name='ConnectorSection', translationalType=JOIN)
connectorewire_region = wireAssembly.sets['Connector wires Set']
wireAssembly.SectionAssignment(sectionName='ConnectorSection',
region=connectorewire_region)

# Creation of step

# Get access to the step objects by using the import statement. The
# StaticStep() method is used to create a static step which could be used
# for loading. This is the step next to 'Initial' step created by default.
# Enter the initial increment and maximum increment as shown below.

import step
```

```
wireModel.StaticStep(name='Load application', previous='Initial',
initialInc=0.1, maxInc=0.1, nlgeom=ON)

# Field output and history output request left at default values
# This example is complicated in nature for a beginner. So I will leave
# these two guys to return their default values 😊

# Boundary conditions
# The four vertices that have to fixed are identified using the findAt()
# method, a reegion is then defined using which a Displacement boundary
# condition is applied.

boundarycondition_vertices = wireInstance.vertices.findAt(((0.0,0.0,0.0),),
((0.0,100.0,0.0),), ((0.0,0.0,50.0),), ((0.0,100.0,50.0),),)
bc_region = regionToolset.Region(vertices=boundarycondition_vertices)
wireModel.DisplacementBC(name='FixOneSide', createStepName='Initial',
region=bc_region, u1=SET, u2=SET, u3=SET, ur1=UNSET, ur2=UNSET, ur3=UNSET,
amplitude=UNSET, distributionType=UNIFORM, fieldName='', localCsys=None)

# Loading
# Similarly, vertices for loading are identified, a region is created based
# on these vertices and then used for leading at these vertices.

load_vertices = wireInstance.vertices.findAt(((200.0,0.0,0.0),),
((200.0,100.0,0.0),), ((200.0,0.0,50.0),), ((200.0,100.0,50.0),),)
#load_region = regionToolset.Region(vertices=load_vertices)
wireModel.ConcentratedForce(name='Force', createStepName='Load
application', region=(load_vertices,), cf2=-100.0,
distributionType=UNIFORM)

# Creation of mesh

# Get access to the mesh objects by using the import statement. We will use
# the predefined regions for element type definition. B31 elements are used
# in this simulation. Previously defined regions are used for meshing.

import mesh
```

```
mainmesh_elementtype = mesh.ElemType(elemCode=B31, elemLibrary=STANDARD)
wirePart.setElementType(regions=main_region,
elemTypes=(mainmesh_elementtype, ))
linkmesh_elementtype = mesh.ElemType(elemCode=B31, elemLibrary=STANDARD)
linkPart.setElementType(regions=link_region,
elemTypes=(linkmesh_elementtype, ))
# Seeding all edges. Increase the value of 'number' to obtain finer mesh.
# generateMesh() method is used to generate a mesh on the part.

wirePart.seedEdgeByNumber(edges=mainSection_edges, number=4)
linkPart.seedEdgeByNumber(edges=linkSection_edges, number=4)

wirePart.generateMesh()
linkPart.generateMesh()

# Job Creation and running
# Get access to the job objects by using the import statement. The Job()
# method is used to create a job. Make sure that you enter the correct name
# of the model.

import job
mdb.Job(name='SimpleWireFrameJob', model='Frame analysis', type=ANALYSIS,
explicitPrecision=SINGLE, nodalOutputPrecision=SINGLE,
description='Analysis of frame bending',
parallelizationMethodExplicit=DOMAIN, multiprocessingMode=DEFAULT,
numDomains=1, userSubroutine='', numCpus=1, memory=50,
memoryUnits=PERCENTAGE, scratch='', echoPrint=OFF, modelPrint=OFF,
contactPrint=OFF, historyPrint=OFF)

# The submit() method is used for submitting the job for analysis. The
# waitForCompletion() makes ABAQUS wait till the job is fully executed.

mdb.jobs['SimpleWireFrameJob'].submit(consistencyChecking=OFF)
mdb.jobs['SimpleWireFrameJob'].waitForCompletion()

# Postprocessing
# Get access to the visualization objects by using the import statement.
# We save the odb object and path to variables which could be used for
```

```
# visualization. The viewport size can also be set.
import visualization
frame_viewport = session.Viewport(name='Results Viewport')
frame_Odb_Path = 'SimpleWireFrameJob.odb'
odb_object_1 = session.openOdb(name=frame_Odb_Path)
frame_viewport.setValues(displayedObject=odb_object_1)
frame_viewport.odbDisplay.display.setValues(plotState=(DEFORMED, ))
```

7.3 Summary

It feels good to have come to the end of this challenging example, isn't it? If you have understood the example fully and the script's algorithm, then you are doing great. And you can start writing Python scripts on your own.

But don't worry if you feel lost or not able to fully understand this example, line by line. This example is difficult in its own ways. So I would recommend you to split the problem into a number of smaller segments and write scripts for them sequentially. This practice will drastically improve your understanding and accelerate your learning further.

Okay then, it's time for the next example, take a break and get ready!

8. Steady state thermal analysis

Things to learn

※ Mirroring a sketch

※ Applying thermal boundary conditions

※ Write a Python script to carry out a steady state thermal analysis

8.1 Problem description

Fins are widely used in many applications to dissipate the heat that is generated. In this chapter, we will be investigating the steady state thermal analysis of an aluminium heat sink which has been designed for cooling a microprocessor. Thermal conductivity is set to $k = 170$ W/m/K. Two of the faces are set to a constant temperature of 350K. A load in terms of heat flux is defined in one of the faces. Convection and radiation phenomenon are also defined on some of the faces. Refer Figure 8.2 to understand the boundary conditions used in this example. The dimensions are in mm.

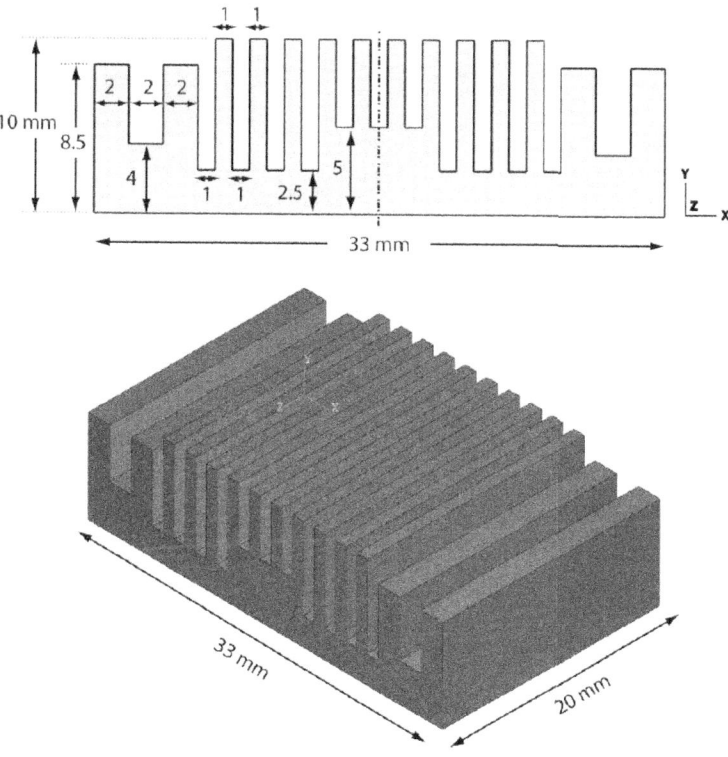

Figure 8.1 Geometry of the fins

Note:

This example and the above figure was taken from
https://www.ccg.msm.cam.ac.uk/images/Module2_Heat_Transfer.pdf

The step by step procedure for doing the analysis in the GUI is explained in the link above.

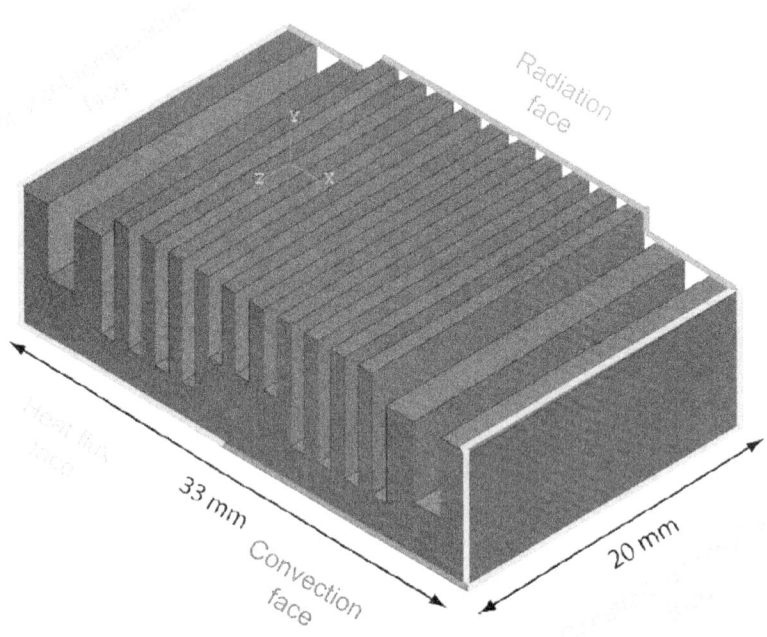

Figure 8.2 Geometry conditions used for this analysis

8.2 Python script

```
# Three dimensional steady state problem
# Heat dissipation through ribbed surfaces
# The first three lines are required to import the required ABAQUS modules
# and create references to the objects that are defined by the module.
# The second line means that you are importing the symbolic constants
# (variables with a constant value) that have been defined by the scripting
# interface of Abaqus. It is a good practice to include these three lines
# at the top of every script that you write. The third line is used for
# acessing the objects of Region() method which is defined inside the
# regionToolset module.
```

```python
from abaqus import*
from abaqusConstants import*
import regionToolset

session.viewports['Viewport: 1'].setValues(displayedObject=None)

# This line is required to make the ABAQUS viewport display nothing.
# 'Viewport: 1' is the default name of an ABAQUS viewport. setValues()
# method is used to set the displayedObject parameter to None so that
# nothing is displayed.

# Model creation

# By default, ABAQUS creates a model named 'Model-1'. We will use the
# changeKey() method to change the name of the model.'mdb' will give access
# to the model database and we will assign it ot 'steadyModel'. The 'One
# main word' strategy for naming variable will start from here. 'steady'
# will be the 'one main word' in this example, for naming main variables.

mdb.models.changeKey(fromName='Model-1', toName='Heat Sink')
steadyModel = mdb.models['Heat Sink']

# Part creation

# These two statements will provide access to all the objects related to
# sketch and part. Including this is not mandatory, but is a good practice.

import sketch
import part

# Drawing one side of the sketch as the other side can be mirrored
# The sketch dimensions are in m.
# In this set of statements, we would define a sketch by using the
# ConstrainedSketch() method. This method in turn has Line() method
# which can be used to features that we want in the
# sketch. Enter a value for 'sheetSize' based on your overall problem size.
```

```python
steadyProfileSketch = steadyModel.ConstrainedSketch(name='Heat Sink Cross
Section', sheetSize=0.2)
steadyProfileSketch.Line(point1=(0.0,0.0), point2=(0.0165,0.0))
steadyProfileSketch.Line(point1=(0.0165,0.0), point2=(0.0165,0.0085))
steadyProfileSketch.Line(point1=(0.0165,0.0085), point2=(0.0145,0.0085))
steadyProfileSketch.Line(point1=(0.0145,0.0085), point2=(0.0145,0.0040))
steadyProfileSketch.Line(point1=(0.0145,0.0040), point2=(0.0125,0.0040))
steadyProfileSketch.Line(point1=(0.0125,0.0040), point2=(0.0125,0.0085))
steadyProfileSketch.Line(point1=(0.0125,0.0085), point2=(0.0105,0.0085))
steadyProfileSketch.Line(point1=(0.0105,0.0085), point2=(0.0105,0.0025))
steadyProfileSketch.Line(point1=(0.0105,0.0025), point2=(0.0095,0.0025))
steadyProfileSketch.Line(point1=(0.0095,0.0025), point2=(0.0095,0.01))
steadyProfileSketch.Line(point1=(0.0095,0.01), point2=(0.0085,0.01))
steadyProfileSketch.Line(point1=(0.0085,0.01), point2=(0.0085,0.0025))
steadyProfileSketch.Line(point1=(0.0085,0.0025), point2=(0.0075,0.0025))
steadyProfileSketch.Line(point1=(0.0075,0.0025), point2=(0.0075,0.01))
steadyProfileSketch.Line(point1=(0.0075,0.01), point2=(0.0065,0.01))
steadyProfileSketch.Line(point1=(0.0065,0.01), point2=(0.0065,0.0025))
steadyProfileSketch.Line(point1=(0.0065,0.0025), point2=(0.0055,0.0025))
steadyProfileSketch.Line(point1=(0.0055,0.0025), point2=(0.0055,0.01))
steadyProfileSketch.Line(point1=(0.0055,0.01), point2=(0.0045,0.01))
steadyProfileSketch.Line(point1=(0.0045,0.01), point2=(0.0045,0.0025))
steadyProfileSketch.Line(point1=(0.0045,0.0025), point2=(0.0035,0.0025))
steadyProfileSketch.Line(point1=(0.0035,0.0025), point2=(0.0035,0.01))
steadyProfileSketch.Line(point1=(0.0035,0.01), point2=(0.0025,0.01))
steadyProfileSketch.Line(point1=(0.0025,0.01), point2=(0.0025,0.005))
steadyProfileSketch.Line(point1=(0.0025,0.005), point2=(0.0015,0.005))
steadyProfileSketch.Line(point1=(0.0015,0.005), point2=(0.0015,0.01))
steadyProfileSketch.Line(point1=(0.0015,0.01), point2=(0.0005,0.01))
steadyProfileSketch.Line(point1=(0.0005,0.01), point2=(0.0005,0.005))
steadyProfileSketch.Line(point1=(0.0005,0.005), point2=(0.0,0.005))

# Line that is used for mirroring

steadyProfileSketch.Line(point1=(0.0,0.0), point2=(0.0,0.005))

# Identify all the lines and store it in a variable
# We use a single alphabet variable here-Not a good practice but acceptable
```

```python
# in this situation though as we will use it a lot.

g = steadyProfileSketch.geometry

# Mirror the lines with respect to the line drawn on the center

steadyProfileSketch.copyMirror(mirrorLine=g[31], objectList=(g[2], g[3],
g[4], g[5], g[6], g[7], g[8], g[9], g[10], g[11], g[12], g[13], g[14],
g[15], g[16], g[17], g[18], g[19], g[20], g[21], g[22], g[23], g[24],
g[25], g[26], g[27], g[28], g[29], g[30]))

# Now that the mirroring would have been done, delete the center line so
# that we would have a closed sketch

steadyProfileSketch.delete(objectList=(g[31], ))

# Using the sketch created above, create a part by extrusion using the
# BaseSolidExtrude() method.

steadyPart=steadyModel.Part(name='Heat Sink', dimensionality=THREE_D,
type=DEFORMABLE_BODY)
steadyPart.BaseSolidExtrude(sketch=steadyProfileSketch, depth=0.020)

# Material creation

# First get access to objects relating to materials by using the import
# statement. Define the name of the material, which would be used further
# in the script. The Density() and Elastic() objects are used to specify
# density and elastic properties as name suggests. The input arguments to
# Density() looks so as it is actually a table with density values with
# respect to temperature. Here we don't need that. So we use blank spaces.
# The same logic applies to Elastic() where it is a table with Young's
# modulus values with respect to Poissons's ratio. We also enter thermal
# conductivity and specific heat.

import material

steadyMaterial = steadyModel.Material(name='Aluminium')
```

```
steadyMaterial.Density(table=((2700, ),        ))
steadyMaterial.Conductivity(table=((170.0, ), ))
steadyMaterial.SpecificHeat(table=((950, ),       ))
```

Section creation and assignment

Get access to the section objects by using the import statement. Then
create a solid section by using the HomogeneousSoldiSection() method. You
can see that we have refered to the material and the profile we created
in the last step. The region is created by using the sequence of cells.

```
import section

steadySection = steadyModel.HomogeneousSolidSection(name='Aluminium
Section', material='Aluminium')
steadyRegion = (steadyPart.cells,)
steadyPart.SectionAssignment(region=steadyRegion, sectionName='Aluminium
Section')
```

Assembly creation

Get access to the assembly objects by using the import statement. The
rootAssembly is an assembly object which is a member of the Model object.
Create an instance of the part by using the Instance() method. By default,
the 'dependent' parameter is set to OFF. Change this to ON. We have
already defined the part name as 'steadyPart'. We will refer to that now.
And do the same for linkPart also.

```
import assembly

steadyAssembly = steadyModel.rootAssembly
steadyInstance = steadyAssembly.Instance(name='Steady Instance',
part=steadyPart, dependent=ON)
```

Datum plane creation and partitioning.
The bottom face has to be partitioned as we have to apply heat flux load
on one face and convection on the other.

```python
steadyPart.DatumPlaneByPrincipalPlane(principalPlane=YZPLANE, offset=0.0)
allCells = steadyPart.cells.getSequenceFromMask(mask=('[#1 ]', ), )
steadyPart.PartitionCellByDatumPlane(datumPlane=steadyPart.datums[3],
cells=allCells)

# Step creation

# Get access to the step objects by using the import statement. The
# StaticStep() method is used to create a static step which could be used
# for loading. This is the step next to 'Initial' step created by default.
# Enter the initial increment and maximum increment as shown below.

import step

steadyModel.HeatTransferStep(name='Heating Step', previous='Initial',
description='Apply heat flux in this step', response=STEADY_STATE,
amplitude=RAMP)

# Boundary conditions - Setting the initial temperature regions
# Refer Figure 8.2 to understand the definitions in the following section.
# The two faces where constant temperature needs to be applied are
# identified.

const_temp_face1_point = (-0.0165, 0.004, 0.01)
const_temp_face2_point = (0.0165, 0.004, 0.01)

# Using the findAt() method, the faces are identified and the regions
# are defined using the regionToolset module.

const_temp_face1 = steadyInstance.faces.findAt((const_temp_face1_point,))
const_temp_face1_region = regionToolset.Region(faces=const_temp_face1)
const_temp_face2 = steadyInstance.faces.findAt((const_temp_face2_point,))
const_temp_face2_region = regionToolset.Region(faces=const_temp_face2)

# Now the steady temperature is set using the TemperatureBC() method.
```

```python
steadyModel.TemperatureBC(name='Constant Temperature Surf1',
createStepName='Heating Step', region=const_temp_face1_region,
distributionType=UNIFORM, fieldName='', magnitude=350.0, amplitude=UNSET)

steadyModel.TemperatureBC(name='Constant Temperature Surf2',
createStepName='Heating Step', region=const_temp_face2_region,
distributionType=UNIFORM, fieldName='', magnitude=350.0, amplitude=UNSET)

# Identify the faces where the heatflux has to be applied and using the
# findAt() method, you can define the faces and then the region.
# The SurfaceHeatFlux() method is used to define the heat flux load.

point_for_heatflux = (-0.0125, 0.0, 0.01)
heatflux_face = steadyInstance.faces.findAt((point_for_heatflux,))
heatflux_face_region = regionToolset.Region(side1Faces=heatflux_face)
steadyModel.SurfaceHeatFlux(name='Heat Flux', createStepName='Heating
Step', region=heatflux_face_region, magnitude=5000.0)

# Define convection and radiation
# Similar procedure is carried out for defining the convection and
# radiation. Also refer to ABAQUS Scripting manual for the correct syntax
# and usage of these methods.

point_for_convection = (0.0125, 0.0, 0.01)
convection_face = steadyInstance.faces.findAt((point_for_convection,))
convection_face_region = regionToolset.Region(side1Faces=convection_face)
steadyModel.FilmCondition(name='Convection', createStepName='Heating Step',
surface=convection_face_region, definition=EMBEDDED_COEFF, filmCoeff=13.0,
filmCoeffAmplitude='', sinkTemperature=200.0, sinkAmplitude='')

radiation_face = steadyInstance.faces.findAt(((-0.0125,0.002,0.02), ),
((0.0125,0.002,0.02), ))
radiation_face_region=regionToolset.Region(side1Faces=radiation_face)
steadyModel.RadiationToAmbient(name='Radiation', createStepName='Heating
Step', surface=radiation_face_region, radiationType=AMBIENT,
distributionType=UNIFORM, field='', emissivity=0.78,
ambientTemperature=320.0, ambientTemperatureAmp='')
```

```
# Absolute zero and Stefan-Boltzmann constant must be set in model
# attributes for problems involving radiation

steadyModel.setValues(absoluteZero=273.15, stefanBoltzmann=5.67E-8)

# Create a set for history output requests and dof monitor

coords_for_set = (-0.0095, 0.01, 0.0)
vertex_for_set = steadyInstance.vertices.findAt((coords_for_set,))
steadyAssembly.Set(vertices=vertex_for_set, name='Set-Node1')

# Field output definition
set_region = steadyAssembly.sets['Set-Node1']
steadyModel.HistoryOutputRequest(name='Required History Output',
createStepName='Heating Step', variables=('NT', ), region=set_region,
sectionPoints=DEFAULT, rebar=EXCLUDE)

# Create a DOF monitor for 'Set-Node1' to monitor throughout the analysis
# DOF 11 corresponds to temperature in ABAQUS

steadyModel.steps['Heating Step'].Monitor(dof=11, node=set_region,
frequency=1)

# Mesh creation
# Get access to the mesh objects by using the import statement. We will use
# the predefined regions for element type definition.
import mesh

allCells = steadyPart.cells.getSequenceFromMask(mask=('[#3 ]', ), )
elemType1 = mesh.ElemType(elemCode=DCC3D8, elemLibrary=STANDARD)
elemType2 = mesh.ElemType(elemCode=DC3D6, elemLibrary=STANDARD)
elemType3 = mesh.ElemType(elemCode=DC3D4, elemLibrary=STANDARD)
pickedRegions =(allCells, )
steadyPart.setElementType(regions=pickedRegions, elemTypes=(elemType1,
elemType2, elemType3))
# Seeding the part. Reduce the value of 'size' to obtain finer mesh.
# generateMesh() method is used to generate a mesh on the part.
```

```python
steadyPart.seedPart(size=0.0005, deviationFactor=0.1, minSizeFactor=0.1)
steadyPart.generateMesh()


# Job creation
# Get access to the job objects by using the import statement. The Job()
# method is used to create a job. Make sure that you enter the correct name
# of the model.

import job

mdb.Job(name='SteadyAnalysisJob', model='Heat Sink', type=ANALYSIS,
description='Heat conduction through Aluminium fins')

# The submit() method is used for submitting the job for analysis. The
# waitForCompletion() makes ABAQUS wait till the job is fully executed.

mdb.jobs['SteadyAnalysisJob'].submit(consistencyChecking=OFF)
mdb.jobs['SteadyAnalysisJob'].waitForCompletion()

# Post processing
# Get access to the visualization objects by using the import statement.
# We save the odb object and path to variables which could be used for
# visualization. The viewport size can also be set.

import visualization

steady_viewport = session.Viewport(name='Steady analysis Viewport')
steady_Odb_Path = 'SteadyAnalysisJob.odb'
odb_object = session.openOdb(name= steady _Odb_Path)
transient_viewport.setValues(displayedObject= odb_object)
steady_viewport.odbDisplay.display.setValues(plotState=(CONTOURS_ON_DEF, ))
```

8.3 Summary

In this chapter, we learnt how to write a Python script for carrying out the a steady state thermal analysis of aluminium fins that are used in a microprocessor. Using this script, you can make minor modifications to carry out a transient analysis as well. We have used the mirroring technique in this example for illustration purposes. But as the geometry is symmetric, you can try the same problem by modelling only one half of the model and then use the symmetry boundary conditions.

I hope you have gained confidence and can write Python scripts independently at this point. Good job so far and keep moving forward!

9. Script for sending email after job completion

Things to learn

※ Monitor a job and send an email based on the job status

※ Exploiting the symmetry of the structure if possible

※ Write a python script that evaluates the stresses in a plate with a hole

※ Importance of indentation while typing Python scripts

9.1 Problem description

The world that we are in right now is full of notifications. I don't know if it is really good. But sometimes, we can use it to our advantage. For instance, if your ABAQUS simulation takes a lot of time to complete, wouldn't it be nice to receive a notification when your job gets completed successfully or gets aborted.

In this chapter, I will teach you to write a Python script which monitors your job and automatically sends an email based on the job's status.

How can there be a book on FEM simulation without the traditional plate with a hole example?

So let's take the famous plate with a hole problem, write a script for carrying out the analysis as well as include the statements that will help us to monitor the job status and send an email after completion.

Figure 9.1 shows the geometry of the plate. As it is symmetric, we will consider only a quarter of the geometry.

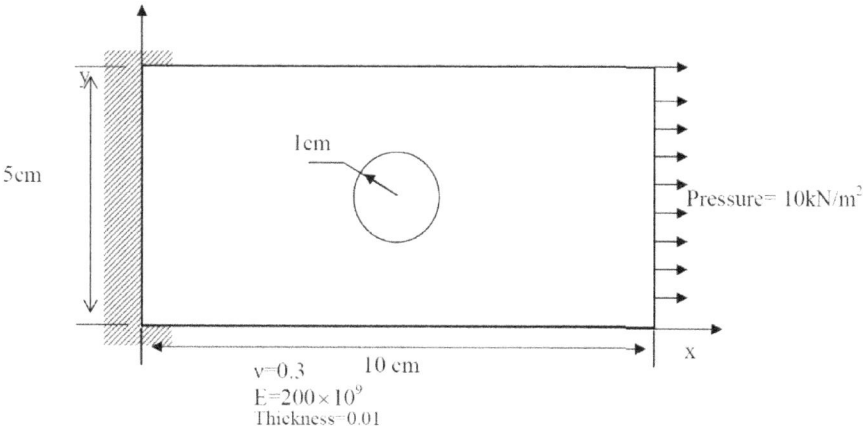

Figure 9.1 Geometry of the plate with a hole

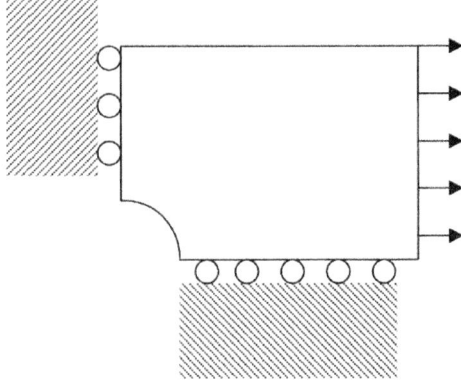

Figure 9.2 Using the symmetry

Note:

This example was taken from

https://sig.ias.edu/files/Abaqus%20tutorial.pdf

The step by step procedure for doing the analysis in the GUI is explained in the above link. The code for the 'email sending' part was inspired from a question and a response in one of the Python communities as well as from Stack overflow.

https://community.esri.com/thread/38595

https://stackoverflow.com/questions/21008030/sending-an-email-with-python-issue

9.2 Python script

```
# Plate with a hole job and send email after job completion

# Write a method called monitorJobCallback() to monitor the job status
# The symbolic constants for messageType can be obtained from the ABAQUS
# scripting manual. Based on the messageType, we will invoke a user defined
# method sendanEmail(email_content)
# We will use monitorManager object provided by ABAQUS.
# removeMessageCallback and addMessageCallback are the methods that are
# used by ABAQUS to specify the function to call and carry out its
# associated task.
```

```python
# if and elif conditional statements are used to check the messageType
# After executing the conditional statements, which means the completion
# or abortion of job, invoke the post processing module and present the
# results.

def monitorJobCallback(jobName, messageType, data, userData):
    if ((messageType==ABORTED) or (messageType==ERROR)):
        # Send an email
        sendanEmail("Oops! - The job has failed")
        monitorManager.removeMessageCallback(jobName='PlatewithholeJob',
messageType=ANY_MESSAGE_TYPE, callback=jobMonitorCallback, userData=None)
    elif (messageType==JOB_COMPLETED):
        sendanEmail("Success! - The job has completed successfully")
        monitorManager.removeMessageCallback(jobName='PlatewithholeJob',
messageType=ANY_MESSAGE_TYPE, callback=jobMonitorCallback, userData=None)
        postProcessing()

# Define a function to send the email
# Notice the indentation used within sendanEmail()

def sendanEmail(email_content):

    import smtplib # Need this for sending emails
    from email.mime.text import MIMEText # Need this to send text in MIME
    # format

    fromAddress = 'example1@gmail.com' # Enter sender gmail id
    toAddress = 'example2@gmail.com' # Enter reciever gmail id
    subject = 'Job monitoring status'
    contents = email_content
    msg = MIMEText(contents)
    msg['Subject'] = subject
    msg['From'] = fromAddress
    msg['To'] = toAddress

    gmail_smtp_server = 'smtp.gmail.com' # Google's Outgoing mail server
    gmail_smtp_port = 587 # Port used b Gmail server for outgoing mail
    gmail_username = '******' # Enter gmail user name
```

```python
    gmail_password = '******' # Type the password during execution
    session = smtplib.SMTP(gmail_smtp_server, gmail_smtp_port)
    session.connect(gmail_smtp_server,465)
    session.ehlo()
    session.starttls()
    session.login(gmail_username, gmail_password)
    session.sendmail(fromAddress, [toAddress], msg.as_string())
    session.close()

# The above 7 statements are very commonly used in most of the Python
# applications. So if you are interested to know deeper about these
# statements, then a google search will easily help you out. If not,
# you can directly copy-paste these statements into your script and modify
# some input arguments to make the script monitor your job and send email.
# The definition of postProcessing() is done here.

def postProcessing():
    import visualization
    hole_viewport = session.Viewport(name='Plate Results Viewport')
    hole_odb_path = 'PlatewithholeJob.odb'
    odb_object = session.openOdb(name=hole_odb_path)
    hole_viewport.setValues(displayedObject=odb_object)
    hole_viewport.odbDisplay.display.setValues(plotState=(CONTOURS_ON_DEF,
))

# The first three lines are required to import the required ABAQUS modules
# and create references to the objects that are defined by the module.
# The second line means that you are importing the symbolic constants
# (variables with a constant value) that have been defined by the scripting
# interface of Abaqus. It is a good practice to include these three lines
# at the top of every script that you write. The third line is used for
# acessing the objects of Region() method which is defined inside the
# regionToolset module.

from abaqus import*
from abaqusConstants import*
import regionToolset
```

```
session.viewports['Viewport: 1'].setValues(displayedObject=None)


# This line is required to make the ABAQUS viewport display nothing.
# 'Viewport: 1' is the default name of an ABAQUS viewport. setValues()
# method is used to set the displayedObject parameter to None so that
# nothing is displayed.


# Model creation


# By default, ABAQUS creates a model named 'Model-1'. We will use the
# changeKey() method to change the name of the model. 'mdb' will give access
# to the model database and we will assign it ot 'holeModel'. The 'One
# main word' strategy for naming variable will start from here. 'hole'
# will be the 'one main word' in this example, for naming main variables.

mdb.models.changeKey(fromName='Model-1', toName='Plate with hole')
holeModel = mdb.models['Plate with hole']

# Part creation
# These two statements will provide access to all the objects related to
# sketch and part. Including this is not mandatory, but is a good practice.

import sketch
import part

# In this set of statements, we would define a sketch by using the
# ConstrainedSketch() method. This method in turn has Line() method
# which can be used to features that we want in the
# sketch. Enter a value for 'sheetSize' based on your overall problem size.

holeSketch = holeModel.ConstrainedSketch(name='Plate Sketch',
sheetSize=15.0)
holeSketch.Line(point1=(1.0, 0.0), point2=(5.0, 0.0))
holeSketch.Line(point1=(5.0, 0.0), point2=(5.0, 2.0))
holeSketch.Line(point1=(5.0, 2.0), point2=(0.0, 2.0))
holeSketch.Line(point1=(0.0, 2.0), point2=(0.0, 1.0))
holeSketch.ArcByCenterEnds(center=(0.0, 0.0), point1=(0.0, 1.0),
point2=(1.0, 0.0), direction=CLOCKWISE)
```

```python
# Using the sketch created above, create a part by extrusion using the
# BaseSolidExtrude() method.

holePart = holeModel.Part(name='Holepart', dimensionality=THREE_D,
type=DEFORMABLE_BODY)
holePart.BaseSolidExtrude(sketch=holeSketch, depth=1.0)

# Material creation
# First get access to objects relating to materials by using the import
# statement. Define the name of the material, which would be used further
# in the script. The Density() and Elastic() objects are used to specify
# density and elastic properties as name suggests. The input arguments to
# Density() looks so as it is actually a table with density values with
# respect to temperature. Here we don't need that. So we use blank spaces.
# The same logic applies to Elastic() where it is a table with Young's
# modulus values with respect to Poissons's ratio. We also enter thermal
# conductivity and specific heat.

import material

holeMaterial = holeModel.Material(name='Generic Steel')
holeMaterial.Elastic(table=((200E9, 0.30), ))

# Section definition and it's assignment

# Get access to the section objects by using the import statement. Then
# create a solid section by using the HomogeneousSoldiSection() method. You
# can see that we have refered to the material and the profile we created
# in the last step. The region is created by using the sequence of cells.

import section

holeSection = holeModel.HomogeneousSolidSection(name='Plate Section',
material ='Generic Steel')
hole_region = (holePart.cells,)
holePart.SectionAssignment(region=hole_region, sectionName='Plate Section')
```

# Assembly creation

Get access to the assembly objects by using the import statement. The
rootAssembly is an assembly object which is a member of the Model object.
Create an instance of the part by using the Instance() method. By default,
the 'dependent' parameter is set to OFF. Change this to ON. We have
defined the part name as 'steadyPart'. We will refer to that now. And do
the same for linkPart also.

import assembly

holeAssembly = holeModel.rootAssembly
holeInstance = holeAssembly.Instance(name='Plate Instance', part=holePart,
dependent=ON)

Step creation

Get access to the step objects by using the import statement. The
StaticStep() method is used to create a static step which could be used
for loading. This is the step next to 'Initial' step created by default.
Enter the initial increment and maximum increment as shown below.

import step

holeModel.StaticStep(name='Load Step', previous='Initial',
description='Apply a pressure load')

Field output requests creation - ABAQUS, by default creates 'F-Output-1'.
We will change the name by using the changeKey() method.

holeModel.fieldOutputRequests.changeKey(fromName='F-Output-1',
toName='Output Stresses and Displacements')
holeModel.fieldOutputRequests['Output Stresses and
Displacements'].setValues(variables=('S', 'UT'))

Application of boundary conditions -
Fix Y for bottom edge and X for left edge -
For doing this, we will first identify the point and use the findAt()

```
# method to find the face and then define the region.

bottomface_point = (3.0,0.0,0.5)
bottomface = holeInstance.faces.findAt((bottomface_point,))
bottomface_region = regionToolset.Region(faces=bottomface)
leftface_point = (0.0,1.5,0.5)
leftface = holeInstance.faces.findAt((leftface_point,))
leftface_region = regionToolset.Region(faces=leftface)


# After definition of regions, we will use the DisplacementBC() to define
# the constraints that replicate the symmetry condition.
holeModel.DisplacementBC(name='FixedY', createStepName='Initial',
region=bottomface_region, u1=UNSET, u2=SET, ur3=UNSET, amplitude=UNSET,
distributionType=UNIFORM, fieldName='', localCsys=None)
holeModel.DisplacementBC(name='FixedX', createStepName='Initial',
region=leftface_region, u1=SET, u2=UNSET, ur3=UNSET, amplitude=UNSET,
distributionType=UNIFORM, fieldName='', localCsys=None)


# Application of load
# Find the load face by a point and then apply the Pressure load

load_face_point = (5.0, 1.0, 0.5)
load_faces = holeInstance.faces.findAt((load_face_point,))
holeAssembly.Surface(side1Faces=load_faces, name='Load Surface')
load_region = holeAssembly.surfaces['Load Surface']

holeModel.Pressure(name='Pressure', createStepName='Load Step',
region=load_region, distributionType=UNIFORM, field='', magnitude=-10000.0,
amplitude=UNSET)


# Meshing
# Get access to the mesh objects by using the import statement. We will use
# the predefined regions for element type definition.
import mesh

elemType_formesh = mesh.ElemType(elemCode=C3D20R, elemLibrary=STANDARD,
kinematicSplit=AVERAGE_STRAIN, secondOrderAccuracy=OFF,
hourglassControl=DEFAULT, distortionControl=DEFAULT)
```

```python
# Select a point inside the plate to find all the cells and then define a
# region based on this sequence of cells.

interior_point = (3.0,1.5,0.5)
holeCells = holePart.cells
allCells = holeCells.findAt((interior_point,))

# Remember, we use the comma while creating a region to indicate that this
# represents a sequence of cells(in this situation)

partRegion = (allCells,)
holePart.setElementType(regions=partRegion, elemTypes=(elemType_formesh,))

# Seeding the part. Reduce the value of 'size' to obtain finer mesh.
# generateMesh() method is used to generate a mesh on the part.

holePart.seedPart(size=0.32, deviationFactor=0.1)
holePart.generateMesh()

# Job creation and running
# Get access to the job objects by using the import statement. The Job()
# method is used to create a job. Make sure that you enter the correct name
# of the model.
# Note that we have used an additional statement here to have the
# functionalities of job monitoring. monitorManager is the ABAQUS object
# which is used for doing this.

import job
from jobMessage import*
mdb.Job(name='PlatewithholeJob', model='Plate with hole', type=ANALYSIS,
description='Job simulates the pressure loading of a plate with hole')
monitorManager.addMessageCallback(jobName='PlatewithholeJob',
messageType=ANY_MESSAGE_TYPE, callback=monitorJobCallback, userData=None)

mdb.jobs['PlatewithholeJob'].submit(consistencyChecking=OFF)
```

9.3 Summary

Firstly, after completing this example, do not worry if you are not able to understand fully – especially the 'email sending' part. There are a lot of help material available in the internet that will make you understand the 'whys' of using these statements. As a ABAQUS user, it is not required to go deeper into those topics unless you are personally interested. The above script can monitor your job and notify you through email based on the job status. If you just want to do that, then you have learnt a very important technique here.

And I am sure that, by this time, you would have fully understood the actual 'plate with hole' part of the script. Can you now notice the general workflow that I have followed in all the Python scripts so far? A lot of statements can be directly reused which will go a long way in increasing your productivity. I am sure that, more you practice this technique, easier it will become to think and solve the problem using Python scripts.

10. Parametrization of a truss

Things to learn

※ The technique to parameterize a truss analysis

※ Evaluate the response of truss under dynamic loading

※ Getting inputs from the user during the analysis

10.1 Problem description

In this chapter, I will teach you the technique of parametrizing a truss analysis using the Python script. Through this example, we will also learn how to get data inputs from the user through the course of the analysis. The truss will be subjected to a dynamic load for a time period of 0.02 s.

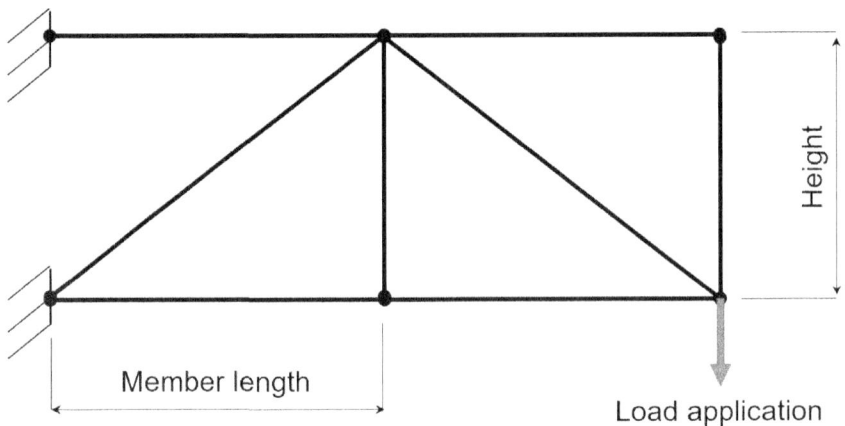

Figure 10.1 Truss under investigation

As you execute the script, ABAQUS will invoke a dialog box prompting you to enter information as shown in Figure 10.2

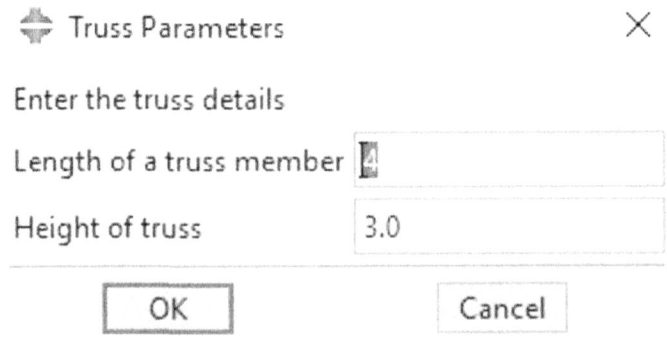

Figure 10.2 A dialog box to enter the truss parameters

10.2 Python script

```
# Parameterization

# The first three lines are required to import the required ABAQUS modules
# and create references to the objects that are defined by the module.
# The second line means that you are importing the symbolic constants
# (variables with a constant value) that have been defined by the scripting
# interface of Abaqus. It is a good practice to include these three lines
# at the top of every script that you write. The third line is used for
# acessing the objects of Region() method which is defined inside the
# regionToolset module.

from abaqus import *
from abaqusConstants import*
import regionToolset

# This line is required to make the ABAQUS viewport display nothing
session.viewports['Viewport: 1'].setValues(displayedObject=None)

# Through the lines below, we will get inputs from the user using
# getInputs() method. This will invoke a dialogbox prompting the user to
# enter data.

input_from_user = getInputs(fields = (('Length of a truss member', '4'),
('Height of truss', '3.0')), label='Enter the truss details',
dialogTitle='Truss Parameters')

# Check if user enters a character insteads of a number. If he enters a
# character by mistake, we will assume some value. Notice the indendation.

try:
    member_length = float(input_from_user[0])
except:
    print 'Unexpected entry. So assuming a length of 4'
    member_length = 4

try:
    truss_height = float(input_from_user[1])
```

```
except:
    print 'Unexpected entry. So assuming a height of 3.0'
    truss_height = 3.0

# Model creation
# By default, ABAQUS creates a model named 'Model-1'. We will use the
# changeKey() method to change the name of the model. 'mdb' will give access
# to the model database and we will assign it ot 'TrussModel'. The 'One
# main word' strategy for naming variable will start from here. 'param'
# will be the 'one main word' in this example, for naming main variables.

mdb.models.changeKey(fromName='Model-1', toName='Truss Model')
paramModel = mdb.models['Truss Model']

# Part creation

# These two statements will provide access to all the objects related to
# sketch and part. Including this is not mandatory, but is a good practice.

import sketch
import part

# In this set of statements, we would define a sketch by using the
# ConstrainedSketch() method. This method in turn has Line() method which
# can be used to draw the lines that we want in the sketch. Enter a value
# for 'sheetSize' based on your overall problem size.

paramSketch = paramModel.ConstrainedSketch(name='2D Truss Sketch',
sheetSize=10.0)
paramSketch.Line(point1=(0, 0), point2=(member_length, 0))
paramSketch.Line(point1=(member_length, 0), point2=(2*member_length, 0))
paramSketch.Line(point1=(0, -truss_height), point2=(member_length, -
truss_height))
paramSketch.Line(point1=(member_length, -truss_height),
point2=(2*member_length, -truss_height))
paramSketch.Line(point1=(0, -truss_height), point2=(member_length, 0))
paramSketch.Line(point1=(member_length, 0), point2=(2*member_length, -
truss_height))
```

```python
paramSketch.Line(point1=(member_length, 0), point2=(member_length, -
truss_height))
paramSketch.Line(point1=(2*member_length, 0), point2=(2*member_length, -
truss_height))

# Using the sketch that we have created, let us create the part using the
# Part() method. TWO_D_PLANAR is an example of symbolic constant.
# The BaseWire() method is used to create a feature object based on the
# sketch we have created.

paramPart = paramModel.Part(name='Truss', dimensionality=TWO_D_PLANAR,
type=DEFORMABLE_BODY)
paramPart.BaseWire(sketch=paramSketch)

# Material creation

# First get access to objects relating to materials by using the import
# statement. Define the name of the material, which would be used further
# in the script. The Density() and Elastic() objects are used to specify
# density and elastic properties as name suggests. The input arguments to
# Density() looks so as it is actually a table with density values with
# respect to temperature. Here we don't need that. So we use blank spaces.
# The same logic applies to Elastic() where it is a table with Young's
# modulus values with respect to Poissons's ratio

import material

paramMaterial = paramModel.Material(name='Steel')
paramMaterial.Density(table=((7800, ),     ))
paramMaterial.Elastic(table=((200E9, 0.29),  ))

# Section creation and assignment

# Get access to the section objects by using the import statement. Then
# create a truss section by using the TrussSection() method. You can see
# that we have refered to the material we created in the last step.
# The next step is to assign the created section to truss members. For this
# we use the findAt() method to find the edges at the provided vertices of
```

```
# the part. With the edges, we can create a region which could be assigned
# to the created section

import section

member_radius = 0.005*member_length
member_area = 3.14*(member_radius**2)
paramSection = paramModel.TrussSection(name='Truss Section',
material='Steel', area=member_area)
truss_section_edges = paramPart.edges.findAt(((member_length/2, 0.0, 0.0),
), ((member_length + member_length/2, 0.0, 0.0), ), ((member_length/2, -
truss_height, 0.0), ), ((member_length + member_length/2, -truss_height,
0.0), ), ((member_length/2, -truss_height/2, 0.0), ), ((member_length +
member_length/2, -truss_height/2, 0.0), ), ((member_length, -
truss_height/2, 0.0), ), ((2*member_length, -truss_height/2, 0.0), ))
truss_region = regionToolset.Region(edges=truss_section_edges)
paramPart.SectionAssignment(region=truss_region, sectionName='Truss
Section')

# Assembly creation

# Get access to the assembly objects by using the import statement. The
# rootAssembly is an assembly object which is a member of the Model object.
# Create an instance of the part by using the Instance() method.By default,
# the 'dependent' parameter is set to OFF. Change this to ON. We have
# defined the part name as 'overhoistPart'. We will refer to that now.

import assembly

paramAssembly = paramModel.rootAssembly
paramInstance = paramAssembly.Instance(name='Truss Instance',
part=paramPart, dependent=ON)

# Step creation

# Get access to the step objects by using the import statement. The
# ExplicitDynamicStep() method is used to create a dynamic step which could
# be used  for loading. This is the step next to 'Initial' step created by
```

```python
# default.

import step

paramModel.ExplicitDynamicsStep(name='Apply Load', previous='Initial',
description='Loads are applied for 0.02s in this step', timePeriod=0.02)

# Apply Loads
# A load, which we will get from user, has to be applied at a vertex. So we
# should identify it using the findAt() method. ConcentratedForce()
# method is used to apply the force at this vertex. Note that, we
# have refered to the step that we created sometime back.

force_input_from_user = getInput(prompt = 'Magnitude of concentrated force
(in -Y direction)', default = '5000')
try:
    force_input = float(force_input_from_user)
except:
    print 'Unexpected entry. So assuming a force of 5000N'
    force_input = 5000

force_vertex_coordinate = (2*member_length, -truss_height, 0.0)
force_vertex = paramInstance.vertices.findAt((force_vertex_coordinate,))
paramModel.ConcentratedForce(name='ForcePulse', createStepName='Apply
Load', region=(force_vertex,), cf2=-force_input, distributionType=UNIFORM,
field='', localCsys=None)

# Apply Boundary conditions

# Similarly the boundary conditions have to be applied at the upper left &
# lower left end of the truss structure. So first we identify the vertices
# and then we apply the boundary conditions at these vertices.

# Pin left end of upper beam

upper_left_end_coords = (0.0, 0.0, 0.0)
upper_left_end_vertex =
paramInstance.vertices.findAt((upper_left_end_coords,))
```

```
paramModel.DisplacementBC(name='Pin1', createStepName='Initial',
region=(upper_left_end_vertex,), u1=SET, u2=SET, ur3=UNSET,
amplitude=UNSET, distributionType=UNIFORM)

# Pin left end of lower beam

lower_left_end_coords = (0.0, -truss_height, 0.0)
lower_left_end_vertex =
paramInstance.vertices.findAt((lower_left_end_coords,))
paramModel.DisplacementBC(name='Pin2', createStepName='Initial',
region=(lower_left_end_vertex,), u1=SET, u2=SET, ur3=UNSET,
amplitude=UNSET, distributionType=UNIFORM)

# Mesh creation

# Get access to the mesh objects by using the import statement. We will use
# the predefined regions for element type definition and for seeding the
# edges. T2D2 is the 2d element type for truss elements. We define the
# mesh size by seeding the edges by a number. This number can be increased
# to have a finer mesh. generateMesh() method is used to mesh the part.

import mesh

mesh_element_type = mesh.ElemType(elemCode=T2D2, elemLibrary=STANDARD)
paramPart.setElementType(regions=truss_region,
elemTypes=(mesh_element_type, ))
paramPart.seedEdgeByNumber(edges=truss_section_edges, number=2)
paramPart.generateMesh()

# Job creation and running

# Get access to the job objects by using the import statement. The Job()
# method is used to create a job. Make sure that you enter the correct name
# of the model. Most of the arguments entered here are not mandatory. You
# can edit the values base don your requirements.

import job
```

```
mdb.Job(name='TrussJobParameterized', model='Truss Model', type=ANALYSIS,
explicitPrecision=SINGLE, nodalOutputPrecision=SINGLE,
description='Analysis of truss under a pulse load',
parallelizationMethodExplicit=DOMAIN, multiprocessingMode=DEFAULT,
numDomains=1, userSubroutine='', numCpus=1, memory=50,
memoryUnits=PERCENTAGE, scratch='', echoPrint=OFF, modelPrint=OFF,
contactPrint=OFF, historyPrint=OFF)

# The submit() method is used for submitting the job for analysis. The
# waitForCompletion() makes ABAQUS wait till the job is fully executed.
mdb.jobs['TrussJobParameterized'].submit(consistencyChecking=OFF)
mdb.jobs['TrussJobParameterized'].waitForCompletion()

# Post processing
# Get access to the visualization objects by using the import statement.
# We save the odb object and path to variables which could be used for
# visualization. The node labels and element labels are turned on for
better # clarity.The viewport size can also be set.

import odbAccess
import visualization
truss_Odb_Path = 'TrussJobParameterized.odb'
odb_object = session.openOdb(name=truss_Odb_Path)

session.viewports['Viewport: 1'].setValues(displayedObject=odb_object)
session.viewports['Viewport: 1'].odbDisplay.display.setValues(plotState=(DEFORMED, ))

# Plot the deformed state of truss

param_viewport = session.Viewport(name='Truss in Deformed State')
param_viewport.setValues(displayedObject=odb_object)
param_viewport.odbDisplay.display.setValues(plotState=(UNDEFORMED, DEFORMED, ))
param_viewport.odbDisplay.commonOptions.setValues(nodeLabels=ON)
param_viewport.odbDisplay.commonOptions.setValues(elemLabels=ON)
param_viewport.setValues(origin=(0.0, 0.0), width=200, height=200)
```

10.3 Summary

As you would have seen, parameterizing is much easier when done through scripting. We took up a truss structure, parameterized the length of individual truss members as well as its height, got this information from the user and evaluated the dynamic response of the truss. The workflow is almost the same and we just have to write the truss dimensions in terms of defined parameters at the necessary places.

I hope, by this time, you would have gotten into the practice of writing your own Python scripts independently. If you face a difficult problem, always try to decompose it into smaller, easily understandable ones and solve them, one by one. A step by step approach will always help you out in these situations. Remember, simpler, the better.

Thank you very much once again for buying my book. My best wishes to you, to learn further and dive deeper into the wonderful world of FEM.

If you have any suggestions for improvement, kindly please do write an email to me at renganathan.sekar@gmail.com.

Good luck!

References

DS Simulia. (2008). *Getting Started With Abaqus, Interactive Edition.* DS Simulia.

Matthes, E. (2016). *Python Crash Course.* San Francisco: no starch press.

About the author

    I have a Master's degree (M.Sc.) in Computer Aided Conception and Production in Mechanical Engineering from the RWTH Aachen University, Germany and a Bachelor's degree (B.E.) in Mechanical engineering from PSG College of Technology, Coimbatore, India.

    I got introduced to FEM as a part of my undergraduate course work 11 years back. Since then my academic interests grew gradually, which made me pursue a Master's degree in Germany with a strong focus on FEM and its applications. And currently I am working with a Korean spinoff in Jinju, South Korea where I apply nonlinear finite element methods to simulate and optimize metal forming processes.

    Other than work, I like to read, remain outdoor and stare aimlessly at nature.

Renganathan Sekar - LinkedIn

Printed in Great Britain
by Amazon